# SONGBIRDS
# IN YOUR GARDEN

Also by John K. Terres

*The Audubon Book of True Nature Stories*

*The Wonders I See*

*Discovery: Great Moments in the Lives of Outstanding Naturalists*

*From Laurel Hill to Siler's Bog: The Walking Adventures of a Naturalist* (Winner of the John Burroughs Medal)

*Living World Books* (editor and contributor)

*The Audubon Society Encyclopedia* (editor in chief and contributor)

*The Audubon Society Encyclopedia of North American Birds* (Winner of a silver medal and citation at the Leipzig Book Fair and the Merit Award of the Art Directors Club of New York)

*How Birds Fly*

*Things Precious and Wild: A Book of Nature Quotations*

*Signals on the Wind: A Book of Animal Behavior*

# SONGBIRDS IN YOUR GARDEN

John K. Terres

Foreword by Roger Tory Peterson

Algonquin Books of Chapel Hill
1994

Published by
ALGONQUIN BOOKS OF CHAPEL HILL
Post Office Box 2225
Chapel Hill, North Carolina 27515-2225

a division of
WORKMAN PUBLISHING COMPANY
708 Broadway
New York, New York 10003


Printed in the United States of America.
Design by Steve Jenkins.



LIBRARY OF CONGRESS CATALOGING-IN-PUBLICATION DATA

Terres, John K.
Songbirds in Your Garden/ by John K. Terres; foreword by Roger Tory Peterson.
p. cm.
Includes bibliographical references and index.
ISBN 1-56512-044-2
1. Bird attracting. 2. Songbirds — United States. I. Title.
QL676.5.T4 1994
598.8 — dc20 94-2277
CIP

10 9 8 7 6 5 4 3 2
First Edition

Cover design: Steve Jenkins
Cover photo: Alan Detrick
Cover illustration of the black-throated blue warbler: John James Audubon
Line illustrations: Margaret S. Nakamura
Bird silhouettes: Irene Metaxatos

# Contents

# Foreword

## Roger Tory Peterson

Feeding birds has been a passion of mine since I was a teenager in high school, well over sixty years ago. In the hills around Jamestown, in western New York State, where I lived, January and February could be very cold; temperatures of 40 below were not unknown. But bitter weather did not keep me from replenishing my feeders thrice a week in the woods north of town, including the very woodlot where a new building will soon house the Institute for the Study of Natural History that has been established in my name.

Setting out on my skis with a knapsack full of seed and suet, it took me about two hours to cover the route from my home to Moon Brook and back. The chickadees and tree sparrows always seemed to be waiting for me. The brash chickadees would actually feed from my hand; the timid tree sparrows would not.

The birds' favorite seed, I recall, was "hemp"—known today as cannabis (or marijuana). The birds seemed to be wild about hemp; they preferred it to millet or cracked corn. Sometimes as I made the rounds I would pop a few of the round kernels into my mouth and chew them. They tasted better than the flat sunflower seeds and didn't require shelling. Reflecting on this, I suspect that those mouthfuls of hemp may have accounted for some of my aberrant behavior.

In Connecticut, where I now live, we have about seventy acres, hardly a suburban garden, nor would I call it an estate; that would be too pretentious. It is basically a rocky ridge covered with oaks, mostly second growth. But from the house and the nearby studio, which are connected by a path through the hemlocks and cedars, I can look from any window and enjoy the birds at the feeders or in the gardens. Lacking water, except on the far side of our property, we had a pump-filled pond excavated outside the studio. Occasionally a kingfisher comes in.

Over a period of thirty years I have spotted about 150 species of birds on our land, nearly 60 of which have nested at some time or other.

The gardens around the house and alongside the studio are planted with shrubs attractive to certain species, and edged with trees that have given haven to robins, wood thrushes, tanagers, orioles, rosebreasted grosbeaks, and several kinds of warblers. Birdhouses, hung from the lower limbs, are quickly taken over by wrens, chickadees, and titmice. But we are too wooded for bluebirds or tree swallows. By leaving the garage door open we have induced phoebes to nest, but in recent years they have refused our invitation, probably because the trees have been closing in. These plain, endearing little flycatchers prefer the semiopen terrain of nearby farms where horses or cattle are their intimates.

We cannot have everything; plant succession dictates what species will live on our property. But we have noticed that during the thirty years of our residence we have gained more birds than we have lost. Because of our program of feeding, gardening, and the erection of nest boxes, several species are much more commonly seen today than they were when we moved in. Cardinals, for example. They were just beginning to penetrate the Connecticut Valley back in the 1950s. Nor were there any tufted titmice. These little gray birds with the jaunty topknots were seldom seen east of the Hudson when I first explored Connecticut. The largesse of sunflower seeds at countless feeders has insured the winter survival of these two southerners with the result that they have now invaded most of New England. So has the mockingbird, but for a different reason. Multiflora rose, an introduced plant considered by some to be a pest, is a winter staple. The juicy red "hips" are small enough for a mocker to swallow. In New England, mockingbirds seem to favor shopping centers, railroad stations, and motels; probably because of the berry-bearing shrubs that are invariably planted to dress up the premises.

Over the years a number of books have been written about

attracting birds, going back to the old classic by Gilbert Trafton published in 1910. I have written introductions or forewords to two or three of them. None of these books has proven more practical or popular than this one by Dr. John Terres, which, in its various editions, has sold more than 500,000 copies since its initial publication in 1953. I suspect that one of the reasons for its wide popularity is its friendly, informal style, spiced with lots of firsthand anecdotes by the author, who has tried out most of the things he describes. It is not a bare bones how-to-do-it book.

Attracting birds, like birding itself, is becoming ever more sophisticated. More people are doing it, and every now and then someone comes up with a novel idea. In this much updated version of his classic, John Terres passes on some of these new ideas for us to try.

Bird-watching, birding, or whatever we call it, has become one of the fastest growing outdoor hobbies. It can be a science, an art, a recreation, a sport, or whatever we choose to make it. It can even be a religious experience as one of my clerical friends insists. And there are many levels, from the watcher at the window to the hard-core elitist who makes lists or who might even travel to the ends of the earth to see new species.

Just how many bird-watchers there are is much debated. Estimates range from two million (conservative) to sixty-three million (according to the U.S. Department of the Interior). The largest category is made up of what might be called the "white-breasted nuthatch type of bird-watcher"—the person who feeds birds at the windowsill or in the garden. It has been estimated that in the United States these good people spend at least two billion dollars a year on birdseed. Everyone up and down our road feeds birds. Some of my neighbors also plant for birds and put out bird boxes. A few have installed birdbaths fed with the garden hose.

This book by Dr. Terres takes us step by step from the window into the garden, be it a suburban backyard, a farm, or an estate in the country. With a little encouragement—let's call it enticement—the local bird population will increase. It might even double or triple.

# Acknowledgments

This new fifth edition of *Songbirds in Your Garden* gives me the opportunity to give special thanks to people who have been particularly helpful with earlier editions—W. L. McAtee, former chief of the Division of Food Habits Research of the U.S. Biological Survey (now the Fish and Wildlife Service), who critically read the manuscript of the first edition; to Dr. Arthur A. Allen of Cornell University for advice on the care and feeding of young birds; to Frank F. Gander in California, who sent me notes from his long experience in attracting birds in that state; to John V. Dennis of Leesburg, Virginia, who sent me lists of birds attracted to feeders throughout the West; to the late T. E. Musselman of Quincy, Illinois, to Bill Duncan and Dick Irwin of Kentucky, William Highhouse of Pennsylvania, and Stiles Thomas of New Jersey, who sent me reports of their work with bluebird trails; to Jim Keighton of the North Carolina Museum of Life and Science for his helpful suggestions; to Ruth Thomas of Morrilton, Arkansas, who sent me some results of her experiments with feeding birds in the garden; to Dr. Douglas A. Lancaster and members of his staff of the Cornell Laboratory of Ornithology at Ithaca, New York; and to the late Dr. Oliver L. Austin, Jr., and his wife, Elizabeth S. Austin, both of the Florida State Museum, for helpful recommendations through the years.

I owe special thanks to Nancy Turel, former librarian of the National Audubon Society, and to her assistant Sema Gurun, for making available long serial lists of ornithological publications that I did not have in my library; to National Audubon's Andrew Bihun, Jr., for many favors and for his suggestions about the care and feeding of young birds drawn from his considerable experience with them; and to Rich Sintchak for California birds that ate Marvel-Meal.

Although their names are too numerous to include, I offer my heartfelt thanks to all those readers of *Songbirds in Your Garden* who have written to me through the years, relating experiences of their own in bird-attracting and offering suggestions for some changes and for proposing, from time to time, new material for inclusion.

My thanks also to Robert Alden Rubin, senior editor, and his assistant, Rob Odom, of Algonquin Books for their superb editing and kindly advice in this latest edition.

And most of all my gratitude to my late wife, Marion, whose keen observations of all living things in our garden and her special love for its flowers, trees, and birds inspired me first of all to write this book. I think she knew how much I appreciated her always valuable advice and practical criticisms. I have tried to show what our garden meant to her in my dedication of this edition.

TO MARION
(1904-1977)

*Who made our garden*
*Tall, with love*
*For all*
*God's creatures,*
*Great and Small*

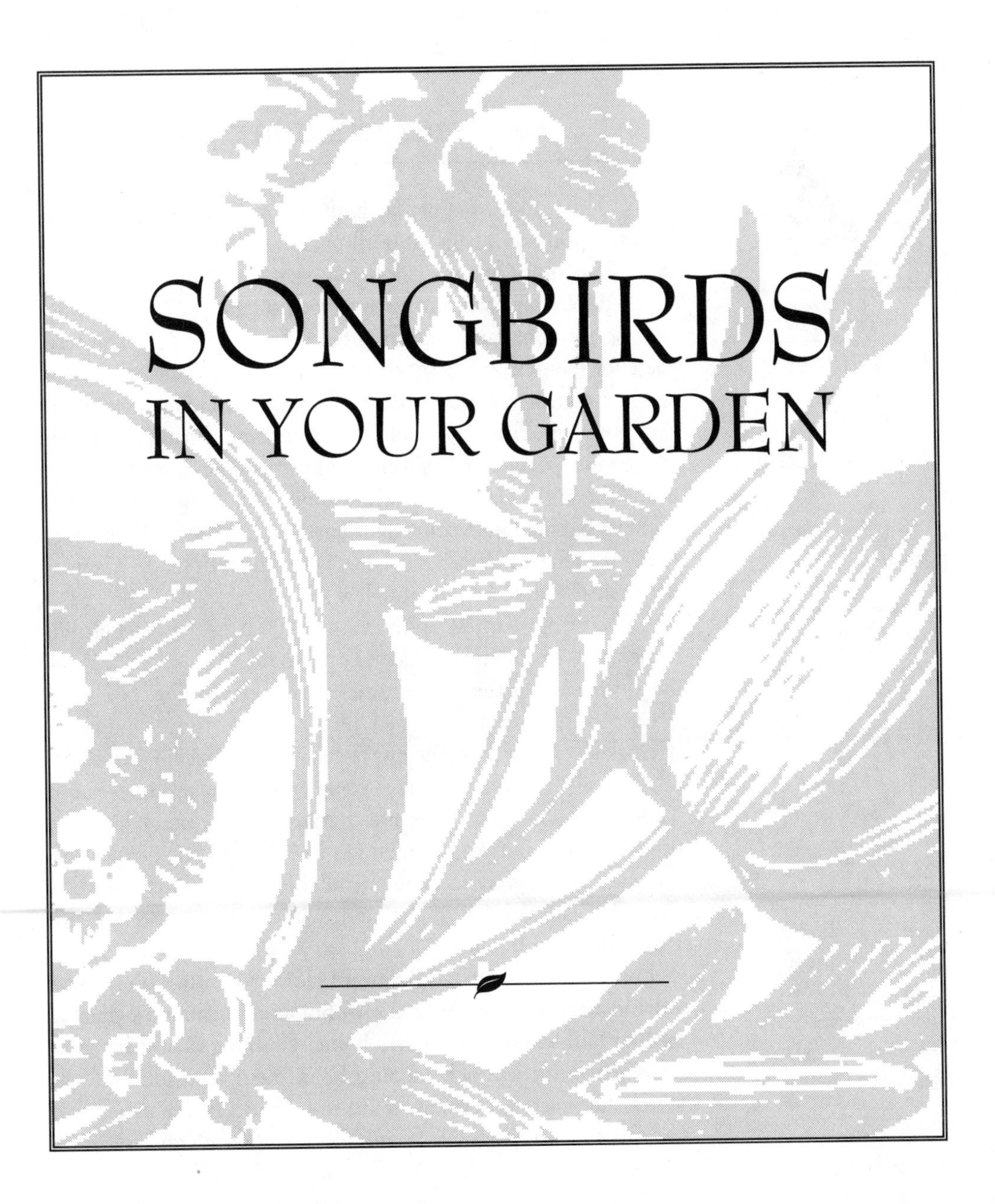

# SONGBIRDS
## IN YOUR GARDEN

**CEDAR WAXWING**

| | |
|---|---|
| Number of Eggs Laid in Clutch: | 3-5 |
| Days to Hatch: | 12-16 |
| Days Young in Nest: | 14-18 |
| Number of Broods Each Year: | Usually 1, sometimes 2 |
| Lifespan: | 3 to 7 years |

*The cedar waxwing ranges from southeastern Alaska to northern California, and across the northern United States to the Atlantic coast and south to Georgia. Its nest is built of twigs, grasses, weeds, mosses, and lichens in the fork or on the horizontal limb of a deciduous or coniferous tree, in an orchard or shade tree, sometimes in colonies in a clump of pines, 6 to 50 feet above the ground. The cedar waxwing may even come to one's hand for string or yarn for nesting material.*

# Introduction

Although my mother fed birds in winter when I was a child, I did not attempt it until many years later. I have a vivid memory of her in a heavy shawl and long skirts that touched the snow, sweeping away a place under a south-facing window of our home and scattering bread crumbs and chicken feed there for the juncos. These were the winter "snowbirds" for which she had a great affection and of which she spoke pityingly when the snows and bitter winds of December came. That was early in this century—the winters were colder then—and, to her, these were "the birds, God's poor, who cannot wait." Later my mother told me that her bird-feeding must have begun about 1910 or 1912, sometime after my father moved us from Philadelphia to Mantua, the country village in southern New Jersey where I grew up.

Some of our earliest Americans fed birds as my mother did. Thoreau, at his cabin by Walden Pond, during the winter of 1845-46, threw out on the hard crusted snow half a bushel of ears of sweet corn that had not ripened. Blue jays came to eat, and flocks of chickadees to pick up the bits of corn left by a feeding red squirrel. A small group of chickadees came to glean their dinner out of Thoreau's woodpile and the bread crumbs he spread at his door, uttering "faint, flitting, lisping notes like the tinkling of icicles in the grass." They were so tame one alighted on an armful of wood that Thoreau was carrying in and, without fear, pecked at the sticks piled near his face.

In the winter of 1876-77, Edward Howe Forbush, who became a distinguished Massachusetts ornithologist, began a program of feeding birds about his country home. He hung suet at his windows and built a rough feeding board near a windowsill, on which he scattered crumbs and grain. From inside a few feet away, the eighteen-year-old naturalist could watch and study the behavior of living birds, as he did for most of his long life.

Later, Forbush and a distinguished colleague, William Brewster of Cambridge, Massachusetts, developed and experimented with types of birdhouses, which Forbush set about his home, even attaching some of them to the outside window frames of his house. Forbush's purpose was not only to closely study birds and to provide for them in winter when natural foods might be locked under snow and ice, but to hold some of them the year round. He knew of their usefulness in eating insects harmful to his woodland, his fruit trees, and his garden crops. (For a more detailed account of Forbush's experiments, see Recommended References in the Appendix.)

**Useful Birds**

**One of the earliest books on what birds eat is Edward Howe Forbush's *Useful Birds and Their Protection,* published in 1908. As an encyclopedic reference, it offers an account of the services of birds in eating insects and rodents destructive to farm crops, includes a full chapter on attracting birds, illustrated with Forbush's own excellent photographs and line drawings, and is still a useful source of information. It is available in some libraries.**

## *Early Books on Attracting Birds*

In 1899, Detrich Lange, professor and instructor of nature study in the public schools of St. Paul, Minnesota, published *Our Native Birds: How to Protect Them and Attract Them to Our Homes*. It was apparently the first book in America devoted almost wholly to attracting birds. The chapter on bird-feeding was written by a guest author, Mrs. Elizabeth B. Davenport of Brattleboro, Vermont, who, in the winter of 1895-96, began methods of feeding birds in her garden such as we use today. She attracted twenty-one species of birds, some of which continued to come for the food she provided and to remain with her the following summer. Of her pioneering efforts, Mrs. Davenport wrote: "I began this work from love of, and companionship of, these feathered friends . . . I know of no other pursuit that brings richer rewards."

In 1910 appeared a book, *Methods of Attracting Birds*, by Gilbert H. Trafton, supervisor of nature study at Passaic, New Jersey. The book was sponsored by the National Association of Audubon Societies (now the National Audubon Society) and was the most complete book on the subject up to that time. Besides chapters on building birdhouses, birdbaths, and winter feeding and planting for birds, it included chapters on bird protection in schools and on bird photography.

One of the most interestingly written bird books of the time, with an introduction by Theodore Roosevelt, former president of the United States, was Ernest Harold Baynes's *Wild Bird Guests*, published in 1915. It not only demonstrated Baynes's love for birds and his devotion to them and to their protection but was filled with the enthusiasm and charm of the writer. Baynes was a former newspaperman, a photographer, a highly popular lecturer, a naturalist, and organizer of some two hundred bird clubs throughout the country. He called his hometown of Meriden, New Hampshire, "The Bird Village." There most of the town's residents attracted birds throughout the year; a town where the birds had been so tamed by the residents that they allowed themselves to be picked up.

"There is no mystery about it," wrote Baynes. "It is simply a matter of being quiet and gentle with your guests and of using a little thought and ingenuity for their welfare." Pine grosbeaks, white-winged crossbills, redpolls, pine siskins, nuthatches, and chickadees were among the wild birds tamed at Meriden.

**Must You Be Able to Identify Birds Before Attracting Them?**

One of the nice things about bird-attracting is its informality. The birds don't care whether you recognize them or not. Unlike the guests you invite to your house for dinner, the birds don't need a formal introduction. Perhaps your purpose in attracting them may be solely to learn their names, by watching them and comparing them so as to sort out the different kinds. You can go ahead confidently with your bird-encouragement program before you are able to recognize even a house sparrow, but it will add a lot to your fun if you know a few of the many kinds that will undoubtedly visit you. Identifying birds will get to be a very exciting game, and it can be sharply competitive if you get your family interested in seeing who can learn a new bird each day.

## *Hand-Feeding Birds*

One severe winter, when pine grosbeaks came down from the north, they were so fearless that Baynes and his wife and other bird-attractors in Meriden could sit down in the middle of a flock and have the birds come to their laps to feed. The grosbeaks even alighted on the heads of children who were watching them and sometimes allowed Baynes to pick them up, one in each hand.

During some snowy winters, when the redpolls and pine siskins came in flocks of hundreds to Meriden, the streets were alive with the birds. They swarmed over the open square of the town and into the yards of the residents. If a person stood still, holding millet or crumbled doughnuts in the extended open hand, palm up, siskins and redpolls alighted there to feed. Baynes found the black-capped chickadee the tamest of all. This is the tiny bird (weighing about one-third of an ounce) that Ralph Waldo Emerson called "a scrap of valor" for its fearlessness and its ability to withstand bitter winter cold.

In the early morning, before Baynes and his wife were up, their chickadees came to the bedroom windowsill, sat there in a row, and hammered on the glass with their small bills to let the Bayneses know they wanted to be fed. When Baynes opened the window, in came the chickadees to eat the crumbled black walnut kernels he had spread for them on the dressing table.

Baynes once invited the chickadees in for breakfast. He and his wife set their table close to the open window and sprinkled bits of walnut kernels on the tablecloth. While the Bayneses were eating, in came the chickadees. They looked up at their hosts, then picked up the nuts and flew off with them into the garden.

One day, while Baynes was living in Stoneham, Massachusetts, he saw a flock of black-capped chickadees in a tree. He held out a handful of broken walnut kernels in his open palm and whistled the "phoebe" call of this bird. Instantly one of them flew down to his hand, picked up a piece of nut, and flew off. On its return trip, Baynes tested its tameness further. He held a piece of walnut between his lips and extended the forefinger of his right hand just in front of his chin as a perch. The chickadee alighted on his finger, leaned forward, and with its bill took the kernel from Baynes's mouth. Mrs. Baynes, at their home in Meriden, once went out into the yard and in her open palm held up broken kernels of English walnuts. Instantly five of the flock of chickadees that wintered in their garden flew down and alighted on her head, shoulders, and extended hand.

**BEWICK'S WREN**

| | |
|---|---|
| Number of Eggs Laid in Clutch: | 5-7 |
| Days to Hatch: | 14 |
| Days Young in Nest: | 14 |
| Number of Broods Each Year: | 2 to 3 |
| Lifespan: | Potentially 3 to 6 years, same as related housewren |

*The Bewick's wren ranges mostly from southern Canada east to central Pennsylvania and Virginia and south to Georgia, South Carolina, and Baja California. Its nest is built in almost any cavity, in knotholes, trees, fence posts, mailboxes, old woodpecker holes, and bird boxes. Nesting materials include mosses, sticks, dead leaves, cotton, hairs, and wool; the inner cup is lined with feathers.*

## What Is a Songbird?

All of the so-called songbirds are classified or grouped by ornithologists (taxonomists) in the order Passeriformes (from Latin, *passer*, sparrow, and *forma*, form—sparrow-shaped). Passeriformes is one of the higher collective categories in bird classification and one of the most stable in the grouping of birds with structural likenesses. Three-fourths of all birds in the world are classified in the Passeriformes.

In North America, Passeriformes includes twenty-nine families of songbirds that are "blood-related," that is, they evolved as a group. In size they range from the largest—ravens, crows, and jays—to the smaller birds we often see in or around our gardens—robins, cardinals, thrushes, mockingbirds, catbirds, starlings, orioles and tanagers, flycatchers (phoebe, for example), swallows, nuthatches, titmice, finches, sparrows, vireos, and warblers.

They all have in common a foot with three toes directed forward and one backward. All toes are on the same plane, or level, are easily movable by the bird, and are adapted to gripping branches of trees, and in the smallest of birds to grasping and perching even on a tiny twig, reed, or stem of grass. Also, the muscles and tendons of the legs are so developed that if the bird has a tendency to fall backwards, they tighten its grip on the perch. Besides "songbirds" they are also called "perching birds."

Another feature that distinguishes the songbirds is that they have several pairs of syringeal muscles that control sounds from the bird's syrinx, or voice box. However, not all songbirds sing—most flycatchers, the house sparrow, and the cedar waxwing, for example. And although crows, starlings, and blue jays are not "singers", they are excellent mimics of the songs and calls of other birds.

Even when Baynes was in the winter woods some distance from his home, sitting in the snow eating his lunch, chickadees flocked about him, alighting on his cap, shoulders, and snowshoes. When he held a sandwich in his mouth, one of the chickadees alighted on the other end and with small bites helped him eat it. One spring day, after being away on a lecture tour of several months, Baynes was delighted on coming home to be met in his lane, half a mile from his house, by a band of his chickadees. While they flew along escorting him home, they alternately flew to him and alighted on his head and shoulders.

**CAROLINA WREN**

| | |
|---|---|
| Number of Eggs Laid in Clutch: | 4-8 |
| Days to Hatch: | 14 |
| Days Young in Nest: | 12-14 |
| Number of Broods Each Year: | 2 |
| Lifespan: | 3 to 6 years |

*The Carolina wren ranges from Nebraska east into southern Ontario and southeastern Massachusetts and south to central Texas and southern Florida. Its nest is built in the cavity of a tree, stump, or hole in a bank, in the upturned roots of a tree, in a stone wall, in old clothes, rafters, or in small baskets in buildings. Its nesting materials include grasses, weeds, leaves, mosses, and feathers. The Carolina wren will also nest in birdhouses.*

# *Living with Songbirds in Your Garden*

If you follow the instructions in this book, you will have opportunities to live as closely with birds as did Baynes and many others. In time, you may have many delightful experiences with individually friendly and tamable birds. But unless you live in winter in snow-covered parts of the northern United States and Canada, you may not have the remarkable experiences that Baynes had with large, fearless flocks of pine grosbeaks, pine siskins, and redpolls. About every two to six years these birds, year-round residents from Canada and Alaska, move southward in fall and winter in great flocks into Maine, Massachusetts, Pennsylvania, and beyond. They are driven south, not by severe winters but from occasional shortages of their natural foods—crop failures of seeds and wild fruits in the forests of the north. Even if you live to the south of these periodic invasions by northern finches, you will have those constant day-to-day bird visitors to your garden that will learn to come to your hands for food. But first, you must start with a winter feeding program.

In our garden at Little Neck, New York, my wife and I had the unending delight and thrill of small, bright-eyed wild birds alighting on our hands for food—black-capped chickadees, white-breasted nuthatches, downy woodpeckers, and, in summer, a catbird, a house wren, and a pair of nesting rose-breasted grosbeaks; during nine years of bird-attracting in North

Carolina, a red-bellied woodpecker, a pine warbler, a ruby-crowned kinglet, and pine siskins. This is not unusual. Some of my friends and correspondents in the northern United States and Canada have hand-fed chestnut-backed chickadees, purple finches, flickers, hairy woodpeckers, the very tame gray jays, evening grosbeaks, hoary redpolls, Bohemian waxwings, starlings, and others. In Florida, I knew of a man who got mockingbirds, a Carolina wren, and tufted titmice on his estate to alight on his hands and feed, and even got a mated pair of tall, wild sandhill cranes to come walking each day to meet him and take food from his hands.

Such are the friendships and intimacies with wild birds that are possible for one who feeds and protects them. And as your garden birds become an increasing part of your life, you will experience a new happiness, a deepening interest, and an enrichment of your daily living beyond your expectations.

It is more than fifty years since I first began to attract birds. In all that time, living with birds in my backyard or garden wherever I have been, I have never known a dull moment. This is the highest recommendation I can offer you for a hobby and an enduring daily occupation that has given me some of the happiest, most absorbing, and contented times of my life.

# Part 1

## *Food and Shelter*

# *Feeding Birds to Attract Them*

ometimes it is purely chance that gives us a new interest that may last a lifetime. Many, many years ago, when I was a government field biologist in upstate New York, I worked with a soil conservation program that helped vastly to increase wildlife on farmlands, but I did not attract songbirds to my own backyard. I had studied birds for many years and I had gone to the woods and the marshes and the open fields to find them. At that time I had not learned that I could bring many of the birds to me, instead of traveling long distances by automobile or afoot to see them.

One day someone told me of a farmer nearby who fed chickadees, nuthatches, tree sparrows, and other native birds. Much interested in the success I heard this man had in taming wild birds, I visited him one cold January morning. I was unprepared for the reception I was to get.

I had walked only partway up his snow-choked driveway when I heard a flutter of wings. Before I realized what was happening, a chickadee, a mite of a bird dressed in black and gray feathers, alighted on my shoulder!

Filled with wonder at this experience, I stood still, not daring to move. The tiny bird looked up into my face, a question clearly showing in its beady black eyes. Suddenly a man spoke.

"Take some of these peanuts. Blacky wants to be fed."

I moved slightly and the chickadee flew from my shoulder to the woolen cap of the ruddy-faced farmer who stood by some shrubbery near his house a few yards away.

"Here!" The man tossed me some shelled peanuts. "Put half a kernel on your lower lip and don't move."

I did as I was told and Blacky the chickadee came flying directly for my face. As the little bird neared me it swooped downward, then up. I felt a slight tickling of my face as the bird's feet touched my chin, the momentary clasp of its claws as it gently pecked the peanut from my lips, and then the light flutter of its wings as it flew

**BLACK-CAPPED CHICKADEE**

| | |
|---|---|
| Number of Eggs Laid in Clutch: | 6-8 |
| Days to Hatch: | 11-13 |
| Days Young in Nest: | 14-18 |
| Number of Broods Each Year: | 2 |
| Lifespan: | 5 to 13 years |

*Black-capped chickadees range across middle and northern North America. A chickadee pair will dig a nest hole in a rotted birch or pine stub usually 1 to 10 feet up. They will also nest in woodpecker holes and birdhouses. Their nests are made of plant fibers and wool.*

up to the branch of a tree to eat its prize.

The farmer's grin told me how astonished I must have looked. It was the first time that I had ever had a wild bird come to me with no more fear than if I were a tree or a bush. I have never forgotten the thrill of that experience and what it taught me about the trust that we can instill in a wild creature by showing it patience, kindness, and understanding.

Since that day I have had many chickadees and other kinds of songbirds come to my hands to feed. Everyone who feeds birds should make it his goal so to gain the trust of at least one wild bird at his feeder, or feeders, that it will come to him, to his children, and to his guests for food. An experience of this kind will win the hearts of more people than all the pleading for bird conservation made in all the bird books ever written.

## WHEN SHOULD YOU START FEEDING BIRDS?

If you start to feed birds in any month of the year, you will probably succeed in attracting them. The best time to begin is in the fall before birds have settled down in their chosen wintering territories and have fixed their habits of searching for food over about the same courses each day. If you set up your feeder early and keep it filled with food you will attract many wintering birds that will become accustomed to visiting it before cold weather begins. During the past 60 years I have lived in many different places, and in each one of those places I have attracted birds: from the Finger Lake region of upstate New York, where the winters were bitterly cold and snow-filled; to Long Island, where the winters were not so severe; to North Carolina, where the temperature in winter seldom falls below 15 degrees Fahrenheit, and where snow, if it comes, usually melts within a day or two.

In the North I began my feeding about October 1; in the South, in late October or early November. The dates of beginning will not matter if you decide to feed birds all the year around, as many of us do.

# Your First Step—
# Putting Up the Bird Feeder

One of the first birds to come to our plain, unpretentious feeder was a house (or "English") sparrow. Before long, other birds—juncos, tree sparrows, and white-throated sparrows—joined the house sparrows where all of these birds fed peaceably, side by side. In all the years of bird-attracting that I have experienced since then, I can always be sure that when we put a feeding station in a new location, house sparrows will lead the other birds to it. Even if house sparrows did not eat Japanese beetles and other insects in my garden, I think I would still welcome them to the feeders for their "guide services" to other birds.

We have been using our open feeder for many years and it has been very satisfactory to the birds. The only change or improvement I made was to build it larger—it is now 36 inches long and 24 inches wide and it accommodates more birds feeding in it at the same time. Many birds, except starlings, house sparrows, red-winged blackbirds, and certain others that flock together, do not like to be crowded closely with others, either of their own kind, or other species, while they are feeding. The larger the feeding tray, the more space they will have in which to feed without fighting among themselves.

## ADVANTAGES OF THE OPEN FEEDER

Although the open feeder has the disadvantage of exposing the food in it to rain and snow, it is so easily made and so attractive to birds that I recommend it highly to every beginner. Later you may want to put a roof over it, or replace it with one of those handsome "roofed over" bird feeders that look like neat little model homes. You can buy these from dealers in bird feeders and

### Building the Open Tray Feeder

If you want your bird guests to be comfortable while eating, and safe from dogs and cats, you will want a table for them. Set it about 4½ feet above the ground on a 2 inch by 4 inch or 4 inch by 4 inch post, which should be set in the ground about 24 to 30 inches deep.

Before you set up your feeding station post, or support, be sure that it will be in place on the open lawn or at the edge of the lawn where you and your family can see it from one of your windows. Be sure that you do not put it directly under the low-spreading limbs of a tree or bush from which squirrels can jump down upon it. We feed the squirrels but we don't allow them on our feeders, where they keep birds from feeding for long periods in cold and snowy weather. Be certain, also, that your location for the feeder is in a place somewhat protected from cold winter winds, perhaps on the south side of your house, and within 10 to 20 feet of a clump of evergreens or other shrubbery into which small birds can fly to escape from hawks and other creatures that occasionally feed on songbirds.

After you have decided on the location of your feeder, you may want to make your own feeding tray, or "table," that you will set for your bird guests. My first feeding station, built many years ago, was a simple open tray. Out of boards ⅞ inch thick I made a shallow rectangular box 18 inches long by 12 inches wide, and about 2 inches deep. In the bottom of it I bored half a dozen holes 1 inch in diameter to allow rainwater or melting snow to drain out of the feeder. Although the holes would also permit some of the seeds or grain to drop to the ground, I didn't worry about their going to waste. The sharp-eyed birds would find the seeds quickly and would enjoy eating them just as much on the ground or in the lawn grass as in the feeder.

The tray I nailed to the top of a post, which, after I had set it a couple of feet deep in the ground, supported the feeding tray at a height of about 4½ feet above the lawn.

bird nesting boxes, but I suggest that you use the open feeder at first until the birds get into the habit of coming to it. In my experience, the birds see the food more easily in the open tray and will learn sooner to come to your feeding station than if it were covered with a roof. This is a very important point to observe because many people have telephoned me or written to say, "Why is it that birds won't come to my feeding station? I keep it supplied with food and yet the birds never come near it."

Usually I find that these beginners fail to attract birds because they use a roofed feeder at the *start*, instead of an open one. Birds either do not see the fine seeds within it or may be too cautious and timid in the beginning to venture under the roof of the feeder. Later, when they become accustomed to coming to your open feeding station, many of them will not object to the roofed feeder and will be quite at home in it. Every year a small flock of white-throated sparrows, a bird that in the eastern states nests from New England and the Adirondack Mountains northward, spent the winter in our backyard in suburban New York City. One of these white-throated sparrows (I think it was the same one) during the winter of 1951-52 spent a lot of time in a roofed weather vane feeder I had set up just outside my living room window. Two of the sides of this feeder were made of wood, the third of glass, and the fourth side was open. Often this sparrow, after having eaten his fill of the small seeds that he always found there, sat in the open side of it for fifteen minutes or more, gazing drowsily around and sometimes closing his eyes to sleep for a few moments. If others of his kind flew into the feeder they would rudely shoulder him aside despite his opened beak and protests at being disturbed.

RULE OF THUMB

**Don't install a roof over your open feeder until the birds get into the habit of coming to it.**

## *Do Birds Taste Their Food?*

Since the first edition of *Songbirds in Your Garden*, many people have asked me if birds like or dislike the taste of some foods, just as we do. I still do not know, as the sense of taste is probably the most difficult of all to test because it is subjective, and birds cannot talk to tell us whether they like the taste of certain

### The Weather Vane Feeder

Use boards 3/4 inch or 7/8 inch thick. Make the feeder 24 inches wide and about 20 inches deep. Make it 10 inches high at the front, about 8 inches high at the back. Cut the vanes or "arms" to extend at least 20 inches in front of the feeder. These vanes turn the feeder away from the wind. Slope the sides of the feeder about 2 inches downward, from the front to the back, on which to nail the roof. The top of the dowel in the feeder (see sketch at center and at left) rotates in a round hole slightly larger than the dowel. Also make a hole through the bottom of the feeder a little larger than the dowel so the rotating feeder does not bind against the dowel. Attach a wooden strip about 1½ inches high across the front of the feeder (where the bird is sitting) to prevent seeds from being blown out on the ground.

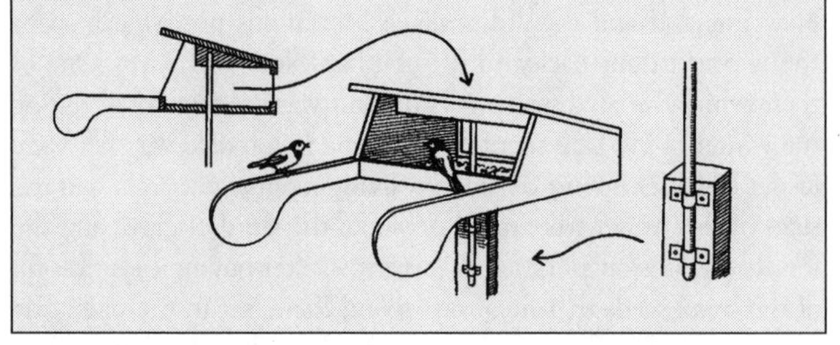

**PAINTED BUNTING**

| | |
|---|---|
| Number of Eggs Laid in Clutch: | 3-5 |
| Days to Hatch: | 11-12 |
| Days Young in Nest: | 12-14 |
| Number of Broods Each Year: | 2, 3, sometimes 4 |
| Lifespan: | Up to 12 years |

*Painted buntings range throughout the southern United States. Their grassy or weedy nests are cup shaped and usually built 3 to 6 feet up in vines, a dense bush, or a low tree.*

foods or simply bolt them because they are hungry.

A striking feature of a bird's mouth is its small number of taste buds, ovoid clusters of chemical sensory receptors that are mostly in the soft area at the base of a bird's tongue, although parrots with a fleshy tongue have them on the tongue itself. Each of us has about 9,000 taste buds; a pigeon has 27 to 59; a day-old chicken has 8, a three-month-old cockerel, 24; some parrots have about 400.

According to one scientist, many experiments with pigeons and chickens suggest that both have an apparently well-developed sense of taste, and these birds markedly rejected a surprising variety of substances. He concluded from these tests and from others that these birds are able to detect at least some substances that taste salty, sour, and bitter; in general, however, he thought that taste plays only a small part when a bird selects its food.

Verne E. Davison, a biologist of the Soil Conservation Service in the southeastern United States, came to different conclusions about the tasting ability of birds. For four years he made backyard feeding experiments with forty-three species of wild, free-ranging North American songbirds, giving them a wide choice of natural foods. He concluded (see *Attracting Birds from the Prairies to the Atlantic*, T. Y. Crowell Company, 1967) that the songbirds in his garden selected their foods mainly on the basis of taste and that color, shape, and texture of the foods were not significant to birds in their choices, under the conditions of his experiments. Size was a factor in a bird's choice when the seed or fruit was too large for it to swallow.

The late Dr. Austin L. Rand, chief curator of zoology, Field Museum of Natural History, Chicago, experimented with the taste and learning of young curve-billed thrashers that he raised and fed by hand. He first offered them the white of hard-boiled eggs cut into small squares, which the thrashers liked. Then he soaked some of the squares in formalin, which, to Rand, was very distasteful. The birds ate some of the squares but then, for a week afterward, refused to eat them. Rand wrote: "They had quickly learned to avoid the ill-tasting food."

From Rand's experiment one might conclude that the proof of the pudding for some birds is indeed in the eating. However, if the birds in our gardens eat a food avidly, with an apparent liking for it, the debate as to whether birds do or do not taste their food may seem to many of us an academic one.

## WHAT KINDS OF BIRD FOOD SHOULD YOU USE?

Although bread does not have a high nutritional value for birds, many of them like it, perhaps for the salt it contains. I

## Experimental Bird Cafeteria

I believe that the *best* seed mixture is the one that you let the birds in your own garden decide upon for themselves. I built a rectangular frame of wood, the same size as the inside of our open feeder. It had eight compartments and looked like one of those frames that separate the ice cubes in your refrigerator. After I had dropped the frame into our open feeder, I filled each compartment with a different kind of seed. For months we kept a record of the seeds that were most eaten and in that way *let the birds tell us* what kind of a small-seed mixture they liked best. Afterwards, instead of buying a ready-made birdseed mixture, we bought our seeds separately and mixed them ourselves. You may not wish to do this, but it repaid us, and was a fascinating experiment besides. The birds in our garden got the seeds they liked and there was little waste because we did not offer them seeds that they wouldn't eat.

Until you have discovered what seed mixture is the best one for your backyard birds, you can attract them by putting baby-chick feed and scratch grain in your feeder. A seed formula my wife and I used in our suburban New York City garden attracted juncos, white-throated sparrows, fox sparrows, song sparrows, tree sparrows, red-winged blackbirds, grackles, cowbirds, and other seed-eating birds. Our mixture contained about equal parts of the small- and the large-seeded millets, white millet, buckwheat, and finely cracked corn. If we wanted chickadees and nuthatches to come to our open feeder, we added sunflower seeds, but we preferred to put our sunflower seeds in a special feeder where the small chickadees and nuthatches could feed undisturbed by the larger birds.

find the greatest value of bread is its conspicuousness in the feeders. Birds see it from a long way off, particularly white bread, and it attracts their attention when finer seeds might not. Blue jays are fond of bread and one of them with a twisted, crippled leg came to our feeders for it all one winter. Blue jays feed very early in the morning, before most other birds. Our crippled jay grew so tame that if I had not put bread in the feeders the night before, early the next morning he would fly to the top of our grape arbor near our back door and watch for me. If I did not come out when he seemed to think I should, he called sharply to attract my attention.

RULE OF THUMB

Blue jays tend to arrive at feeders very early in the morning, before most other birds.

Robins, grackles, red-winged blackbirds, tree sparrows, juncos, cardinals, chickadees, and many other birds like bread. When I put the bread in the feeders, or scatter it on the ground, I break it up into pieces about the size of my fingertips so the birds can eat it easily.

After you start feeding birds, you will undoubtedly have some of those big, black-and-purple grackles coming to your feeder for bread. You will be amused as we always are at their wisdom. If the bread we put out becomes too dry and hard for them to swallow, they will fly with the pieces, one at a time, to our birdbath and dunk them until the bread is soft. Then they will bolt the pieces down and look around at us with what we imagine is a look of self-satisfaction.

TRY THIS

For a special treat, feed your robins coarsely ground toast.

Robins are especially fond of toast. We found that our robins preferred their toast ground up, and so we humored them by running it through a meat grinder before serving it to them.

Although I believe it is best to attract birds largely with seeds, fruits, nuts, and other foods that they are used to eating in the wild, they sometimes develop surprising tastes for certain foods that we humans like. When I was a boy I raised an orphaned robin that had a mania for ice cream! At night we kept him in a canary cage and he never failed to awaken when one of us came home from the store with ice cream. Even though I had covered his cage with a cloth to keep drafts off him while he slept, he chirruped excitedly when he heard the scraping of our spoons in the dishes. After I had teased him by making him beg, I fed him some of my ice cream out of my own dish. With each mouthful this lusty young robin closed his eyes and shivered, but he always opened his mouth and yelled for more as long as I would offer it to him.

I do not recommend that you try to attract birds with ice cream. There are other kinds of foods that are cheaper and that birds will like better.

**BUSHTIT**

| | |
|---|---|
| Number of Eggs Laid in Clutch: | 5-13 |
| Days to Hatch: | 12 |
| Days Young in Nest: | 15-18 |
| Number of Broods Each Year: | 1, possibly 2 |

*Bushtits range throughout the western United States. Their nests are gourd-shaped pockets made of twigs and mosses, which are suspended from a tree or bush 6 to 25 feet in the air.*

# *The Trolley Feeding Station*

Early in my experience in attracting birds, I heard a story that I have never forgotten. A woman who had spent many happy hours watching birds in her village garden developed a heart ailment that put her to bed in an upstairs room for a long rest. Far more worrisome to her than confinement, she fretted over not being able to see her garden birds.

Her son, anxious to help her, talked with a man in a nearby town who attracted birds regularly to his backyard. Between them they built a feeding shelf for the birds, which they attached to the sill just outside the ill woman's window. For a week they kept it well supplied with food, but no birds came to it because the food was apparently too high above the ground for them to see it.

Then the two men built a trolley feeder by first stringing a "running wire," or a lightweight cable from the base of a large tree in the garden to the sill of the convalescent's window. Next, they built an open feeder, and to the top of it, they fastened two small pulleys. The two pulley wheels fitted nicely over the wire so that the feeder now had a track to run on. To the feeder they fastened a long wire, which they ran to the bedroom window from which the feeder could now be drawn along the running wire. All was ready now for their experiment.

After they had filled the feeding tray with bread, sunflower seeds, and a birdseed mixture, they left the feeder suspended about

### The Trolley Feeder

Build your trolley feeder about 12 inches high, from the feeding shelf floor to the peak of the roof; about 18 inches long; and about 12 inches wide. Cut end pieces so that the roof slopes about 2 inches downward. Fasten metal trolleys to the roof, or to the uprights if you are building an open feeder. Use a longer trolley on the uphill side to level the feeder. The seed hopper and the wire suet cage are attached to the wooden uprights. Put a nailing strip about 1½ inches high along the edges of the tray to keep seeds from blowing away.

four feet above an open area in the center of the garden. After a small group of birds were coming to it regularly, they drew the feeder along the ascending wire, a few feet each day. The birds followed it because it was never far from the place where they had been feeding in it the day before. Within a short time the birds had accompanied the feeder in its upward progress all the way to the upstairs window. There the birds came every day to bring happiness to the woman who had been starved for the sight of them.

I know people who have not had to use a trolley feeder to get birds to come to their window shelf feeders, but some of us need it for this purpose. The best way to discover whether or not you need the trolley feeder is first to try feeding the birds at your windowsill. If they don't come there after a week or ten days, you probably need it to do the trick. After using the trolley feeder to get the birds in

**Should You Discriminate Between Your Bird Guests?**

Some people object to feeding the larger and more aggressive birds, claiming that they frighten away smaller ones. In all the years we have fed birds we have never seen the jays, starlings, and grackles in our garden keep the smaller birds from the feeders for very long. If you have only one feeding station you will notice that, as soon as the larger birds leave it, the smaller sparrows and juncos will fly to it and feed there until the larger birds return. We often scattered some grain and pieces of bread on the ground at several places in the yard, which distributed the birds about, and allowed most of them to eat peaceably in our yard, all at the same time. Many of our smaller birds came to our glassed-in window shelf feeder, where they fed undisturbed because the opening was too narrow to admit a blue jay, grackle, or starling.

the habit of coming to your window, you can set it up on a post in your garden and use it as a fixed feeding station, or you can keep it at the window to serve as your window shelf feeder.

## *Discoveries You Will Make About Birds at Your Window*

Window feeders are very desirable for several reasons. They bring the birds so close to you that you won't need binoculars or field glasses to see them, and the children and all of the family can enjoy the birds in comfort while indoors on many a cold winter day. If you are interested in sketching or photographing birds, they will give you plenty of opportunities for close-up portraits of them. If you are intrigued by the behavior of birds, as you surely will be, you will learn a lot more about the way some birds bully or dominate others, their colors, the details of their feathers, how they feed, and other interesting things that might escape you if the birds were much farther away.

Through feeding them at my window, I discovered that birds shiver from winter cold just as we do. At my window shelf I have watched fox sparrows, song sparrows, juncos, and chickadees tremble violently early on frigid mornings in winter. When I lived in upstate New York, at 30 degrees below zero I saw small birds at my window feeder hold up their feet alternately in their body feathers to keep their feet warm. During these bitterly cold days, I sometimes found the bodies of small birds lying frozen on the snow under trees in woodlands where they had roosted the night before. I knew that they had died of the extreme cold, but I also knew that the cold weather wouldn't have killed them if they had been able to find sufficient food during the day to carry them through the long winter night.

After an experience of this kind, I would hurry home and fill my feeders to their fullest, determined that this should not happen to the birds that were wintering in and around my yard.

## THE WINDOW SHELF FEEDER

One of the most popular (with the birds) window feeders I ever saw was a homemade shelf about 12 inches wide and about

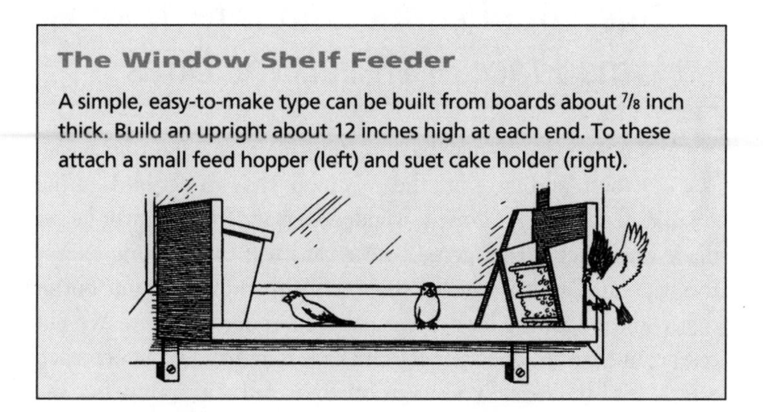

**The Window Shelf Feeder**

A simple, easy-to-make type can be built from boards about 7/8 inch thick. Build an upright about 12 inches high at each end. To these attach a small feed hopper (left) and suet cake holder (right).

36 inches long—the same length as the windowsill. The feeding shelf had a strip of wood about 1½ inches high tacked all around its three outer sides to keep the birdseed mixture from blowing away and was set tight against the windowsill and level with it. Beneath the shelf the owner had nailed two wooden braces, one at each end, which were slanted at a 45-degree angle from the outer edge of the shelf down to the side of the house where nails held them in place.

The day that I visited the owner of this window feeder, I saw purple finches, evening grosbeaks, chickadees, and nuthatches take their turns feeding outside the window within two feet of us while we watched them from the inside through a lace curtain. The curtain permitted us to see the birds, but prevented them from seeing us.

Each end of this window feeding tray had a small feeder hopper, the kind that poultry raisers use in chicken houses, which had been nailed on the feeding shelf. These feeder hoppers, about 18 inches high and the width of the shelf, held several quarts of birdseed, which trickled out of the bottom of the hoppers, as the birds fed upon it. The owner explained that if he wanted to go away for a few days in winter, he filled the hoppers, which assured him that the birds would have enough to eat until he returned.

## *Suet and Peanut Butter to Take the Place of Insects for Birds*

Up to now you may have gotten the impression that bread, toast, and seeds are the only foods that attract birds to the feeders. The downy woodpecker and the slightly larger hairy woodpecker that come to our yard will eat nothing except the beef suet we put out for them, or perhaps the peanut butter that I smear into the bark of the oak tree in our backyard. We put out peanut butter especially for chickadees and the brown creeper, a little bird that usually visits us briefly in fall and spring, but the woodpeckers are more than welcome to it, too. Add cornmeal or suet to peanut butter to prevent the birds from choking, or try recipe for "Marvel-Meal" on page 38.

**WARNING**

Always mix cornmeal or suet into your peanut butter to prevent the birds from choking.

Apparently suet, which chickadees, nuthatches, blue jays, starlings, and other birds also will eat, is a quick source of heat and energy for birds, which makes it particularly valuable to them in winter. It also is a good substitute for the insects that birds usually feed upon but may not find so plentiful in cold weather. Beef suet is now difficult to get from local butcher shops and grocery stores, but you can easily substitute peanut butter mixtures, which have the same nutritional value and actually seem even more attractive to birds.

In a small town where I lived a number of years ago, I had several fine apple trees in my backyard. When I first moved there, I immediately set up my feeders to start attracting birds, which I knew would help protect my apple trees from insects that feed upon them. One of my neighbors, a good man and a "practical" one by reputation, came one day to see what kind of nonsense I had introduced into his community.

At first he looked suspiciously at my neat, brown-painted feeders and questioned me about the cost of the birdseed mixture that he saw in them. Then he noticed a pair of downy woodpeckers methodically pecking over each branch of the apple trees and hitching up and down their trunks. I assured him that the little birds were digging out the white grubs of insect borers, which are destructive to apple and other fruit trees. I did not need to tell him that the stomach contents of hundreds of these birds had been examined by scientists. Nor did I need to say that the results of these bird-food-habit studies had been published in government bulletins to tell people like himself of the kinds of insects that birds feed upon. He had heard about the good that birds do and he nodded his approval of the woodpeckers that were tapping away at my apple trees. "Now that," he agreed emphatically, "*is* a great service to your trees!"

He was surprised when he learned that my woodpeckers were not attracted to my birdseed mixture, but came for beef suet. I showed him the three little wire suet containers, bought from a dealer in bird-attracting equipment, which I had nailed about 5 feet from the ground to the sides of several trees in my backyard. These can be made from hardware cloth or an old wire soap dish.

### The Feeding Stick

These can be made from roughened wood, about 2 inches square and about 2 feet long. Drill l-inch holes through the stick, a few inches apart, and drill a small hole in the top of the stick for a wire loop to suspend it. If you don't want to make your own, you can buy feeding sticks from dealers in bird-attracting supplies.

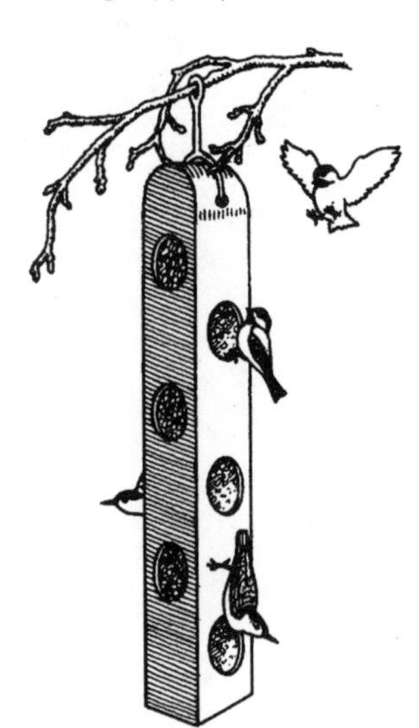

While we watched, the woodpeckers ceased their hunt for the borer grubs momentarily to visit one of our suet feeders. A white-breasted nuthatch fed at another, a chickadee at the third one. Clinging to the sides of the tree trunks, above or below the small wire cages, they pecked through the half-inch wire mesh openings at the firm white beef suet, picking it out in small pieces. I explained to my neighbor that I put the suet in wire cages to keep European starlings, blue jays, grackles, and other of the larger birds that like it from flying away with big chunks of it and leaving the smaller birds without a supply.

At that moment a starling flew down to the most distant suet feeder, alighted on top of the wire cage, and frightened off the little nuthatch that was feeding there. The starling pecked eagerly down through the wire but could bring out only small pieces of the suet. The nuthatch, which it had frightened away, had flown to one of my "feeding sticks" that hung about 5 feet above the ground, suspended by a wire from the branch of a cherry tree. Our feeding stick was a round, 3-inch-thick section of a tree limb about a foot long in which I had bored l-inch-diameter holes and had filled with suet. The nuthatch now clung to this stick and fed quietly without being disturbed by the starling, which had not yet learned to cling to a swaying perch.

My neighbor, apparently impressed with the success of my bird-attracting program, went home and started one of his own. Later he proudly showed me downy woodpeckers and nuthatches feeding on the suet he had put out for them in his trees. He thought, even through that first summer, that the birds had helped him get a better crop of fruit. We are sure that they help to keep our fruit and shade trees and our garden shrubs healthy. Our most welcome bird guests are the woodpeckers, nuthatches, chickadees, creepers, and others that, besides eating our suet, spend most of their time eating the insects in our garden.

TRY THIS

If you are worried about larger birds carrying off too much of your suet and leaving the smaller birds with nothing, try putting out the suet in small wire cages.

## *How Many Feeders Should You Have?*

During the 1920s, near Ithaca, New York, a bird scientist studied the sizes of the wintering territories and the feeding habits of black-capped chickadees and white-breasted nuthatches. From the results of his careful work, we know today exactly how many feeders you need to take care of all the nuthatches and chickadees that will winter in and around your backyard.

You need just one feeder.

One feeder would attract and feed all of the twenty-five or thirty chickadees that the bird scientist discovered would winter on eighty acres of partly wooded country in upstate New York. One feeder would take care of the needs of the one pair of white-breasted nuthatches that lived on each twenty-five acres of woodland near Ithaca. Therefore, one feeder easily accommodated our one pair, or rarely the two pairs of white-breasted nuthatches that came to our quarter-of-an-acre yard on Long Island. It also easily supplied enough food for the three to a dozen chickadees that we fed every winter.

But we attracted far more birds than a few nuthatches and chickadees. We were also host and hostess in winter to blue jays, grackles, cowbirds, juncos, tree sparrows, white-throated sparrows, song sparrows, starlings, house sparrows, goldfinches, and others—about fifty to one hundred individuals of twelve to fifteen different kinds of birds each day. Our backyard and our free lunch were open to all and so we needed more than one feeder to provide for those small, shy birds that would not come to a feeder crowded with larger birds.

All birds need food frequently in cold weather to keep alive, especially when the temperature is at freezing or lower. You probably will have only one feeder in the beginning. Later you may want several to prevent your chickadees, song sparrows, and other small birds from standing by on a cold, snowy day while your lone feeder swarms with a scrambling mob of starlings, blackbirds, house sparrows, and others of the more aggressive birds.

## THE NUMBER OF FEEDERS IN OUR YARD

Even though you may be eager to attract as many birds as possible, you should not have too many conspicuous grain feeders in your yard or it will be what some people describe as "junky." Space your feeders at least 30 feet apart and set them up temporarily until you can look them over to see that they are neat and appropriate in their places and can be seen from one of your windows.

**RULE OF THUMB**

When you first set up your feeders, space them at least 30 feet apart in temporary spots until you are satisfied with their appearance and are sure they are within sight of your windows.

Our yard from the rear of our house to the back property line is about 80 feet deep and 60 feet wide. On the windowsill of one of our living room windows, overlooking the garden, we have a window shelf feeder. Beyond it, in the middle of the lawn within view of this window, we have set up our large open feeder on a 4-foot post. At both rear corners of the garden we have roofed or partly enclosed feeders, set on cedar posts. In these four feeders we always have a supply of our birdseed mixture, some grit, and small pieces of bread or toast, which most birds like.

On the trunks of three trees that are at least 30 to 40 feet apart we nailed suet cages that we keep filled with white beef suet, which birds like best. We had two feeding sticks hung from the branches of trees growing in opposite corners of the garden. We filled the 1-inch holes of these feeding sticks with beef suet and a peanut butter mixture.

From a branch of a small cherry tree we had suspended a

green-painted metal feeder, which we kept filled with sunflower seeds only. These were for the chickadees, goldfinches, purple finches, and the occasional flocks of evening grosbeaks that visited us in winter and preferred sunflower seeds to all other kinds of food.

In an apple tree I hung a wooden feeder, which we also kept filled with sunflower seeds. All of our feeders, which were suspended by heavy wire from the branches of trees, were at least 5 feet above the ground. You should always hang your suspended feeders this high to be out of reach of cats and dogs that will try to catch birds on feeders any nearer the ground. Be careful, too, that you do not hang it any closer than about 8 feet from a fence or a post from which a cat or a squirrel can jump to the feeder.

RULE OF THUMB

Hang your feeders with heavy wire from the branches of trees at least 5 feet above the ground and at least 8 feet from any nearby fences or posts.

## *The Cost of Bird Foods and the Amounts Birds Will Eat*

I have hesitated to mention the costs of bird foods because, like the prices of human foods, they may vary considerably from year to year. One winter some years ago was a season of relatively high prices. If I tell you what it cost us to feed our birds, you will have an idea of the greatest expense that it might be to you. Although we feed birds all the year round, we consider our "winter feeding" to last from about October 1 to April 20 of the following spring. During that time, in 1951-52, the birds in our yard ate 200 pounds of our mixed small birdseeds at a cost to us of 7 1/2 cents a pound, or $15. They also ate 1 pound of white beef suet a week, or 30 pounds for the winter. At 10 cents a pound that we paid for suet, the 30 pounds cost us $3. Our total feeding bill, to keep our yard alive all winter with at least fifty birds each day, was exactly $18. We did not add the cost of bread or toast to our winter feed bill because this is usually stale leftovers that would be thrown away if we didn't feed it to the birds. Of course, it will cost you much more to feed birds today, but you will still find that it is not very expensive.

## Sunflower Seeds and Your "Extra-Dividend" Birds

**HOUSE FINCH**

| | |
|---|---|
| Number of Eggs Laid in Clutch: | 2-6 |
| Days to Hatch: | 12-16 |
| Days Young in Nest: | 11-19 |
| Number of Broods Each Year: | 2, sometimes 3 |
| Lifespan: | 6 to 10 years |

*House finches range throughout the West and also in the eastern United States. Their grassy nests are found in the holes of trees and poles, and in shrubs and vines; they will also nest in birdhouses.*

In late fall or early winter if you one day get a flock of the beautiful, pale-gold evening grosbeaks or the scarlet and rose-colored pine grosbeaks at your feeding station, rejoice in your good fortune. These and the crossbills, birds with beaks oddly crossed like a pair of scissors, are beautiful strangers that usually nest in the North Country, from the Adirondack Mountains of New York far into the Canadian wilderness. They come down to us only in winter.

My wife and I called these birds our "extra dividends" because we didn't get them regularly every winter, nor did they come to everyone's feeding station. To get these extra-dividend birds we invested a little more money in a special food for them. These northern grosbeaks, the crossbills, and the purple finches that sometimes stopped with us on their way farther south were especially fond of sunflower seeds. The evening grosbeaks would not come to our feeding stations unless we had these seeds available for them, but sunflower seeds were expensive.

In 1993, sunflower seeds at my local farm supply stores were between 30 and 50 cents a pound, depending on the quantity bought; birdseed mixtures were about 50 cents a pound—but write to dealers in bird-attracting supplies to get their catalog prices. Many grocery stores no longer sell beef suet for birds, though some mail-order catalogs do (in 1993, the Audubon Workshop charged $21.95, plus shipping, for twelve 1-pound calcium and suet blocks).

Besides the large flocks of evening grosbeaks, purple finches, and crossbills that may visit you and eat 100 pounds or more of sunflower seeds in one winter, redbirds (or cardinals), blue jays, chickadees, tufted titmice, nuthatches, goldfinches, grackles, and even red-winged blackbirds, as we discovered, will eat them.

Fortunately, some of the wild-bird seed mixtures that you can buy fairly reasonably from local feed stores or from dealers in bird-attracting equipment contain from 15 to 25 percent of sunflower seeds. These will take care of most of the birds that eat them and ordinarily winter in your backyard. We liked to have an extra quantity on hand just to feed those beautiful evening grosbeaks if a flock of them suddenly swept down into our garden. We knew they would stay for a while, *if* we kept our feeders filled with sunflower seeds every day.

**TRY THIS**

**You might try growing your own sunflowers to supplement your sunflower seed supply.**

If you have a sunny, open area in your yard, you may wish to grow your own sunflowers. They are not difficult to raise—your county agricultural extension agent can offer advice—and the seeds will sprout quickly in almost any good garden soil. We have raised ours by planting several rows of seeds along our backyard fence. When the heavy-stalked plants grow up ten or twelve feet tall, their big, green leaves and large golden flower heads, a foot or more across, make a colorful and effective background for other kinds of flowers. Unfortunately in a small yard the few rows of sunflower plants that you will have room for will only supplement your sunflower seed supply. It requires about half an acre of ground to raise several hundred pounds of sunflower seeds.

## Some Western Birds in a California Garden

I spent a cool December day years ago with my brother in his San Diego garden and watched the birds that came to the flat board he had fixed on the low horizontal branch of a plum tree. A California scrub jay came almost constantly to the feeding station and was so tame that it freqently came to a birdseed mixture in my brother's outstretched open palm to select out a sunflower seed, then fly with it to a bush to crack open the hull and devour the seed. It then was back in seconds for another. I myself had the pleasure of it coming to my hands with equal trust and boldness.

While we watched, dozens of house finches—the so-called California linnet, harking back to the days when it was popular as a cage bird—and some American and green-backed goldfinches came to the feeder in the plum tree. On the ground, a lone brown towhee in the company of golden-crowned, white-crowned, and white-throated sparrows, mourning doves, Oregon "dark-eyed" juncos, and some Brewer's blackbirds fed on the birdseed mixture my brother had spread on the ground near some bushes in the lower part of his garden.

During the first week of my visit, we watched other songbirds come to the garden. Some American robins came to the feeder for bread. A hermit thrush scratched about under the garden bushes with a spotted towhee and some song and chipping sparrows. One day a red-shafted flicker called from a tree, and, as we watched, an Anna's hummingbird and a Costa's hummingbird hovered on droning wings before the violet-blue flowers of a handsome evergreen rosemary bush in the garden.

The last, and most charming, of the birds to arrive on that first day were two of the very small Brewer's sparrows. These birds nest in summer in the deserts of the Great Basin of the Far West in sagebrushes and cacti, where they may outnumber all other birds.

These two came at once to the baby-chick feed made of ground wheat and corn that my brother had spread at the base of a large rock in his garden—a separate feeding place for the smaller birds. From the moment of their arrival, the Brewer's sparrows were extraordinarily tame and allowed me to walk within a few feet of them without showing fear. As a California friend of mine, himself a fine naturalist, wrote, "There is something eternally babyish about this little bird, its tiny form and innocent trusting face like that of a newly hatched chick."

It was all quite a show for me, an easterner, certainly equal to my experience with feeding the birds in winter in my Long Island, New York, garden. And I was excited over some of those western birds I had never seen before and now could add to my lifelong birding list.

## EXTRA-DIVIDEND BIRDS IN MY BROTHER'S CALIFORNIA GARDEN

These were birds that did not come every winter. That day might have even been their first vist to my brother's garden. The brown towhee that we saw—a lone gray-brown bird with a relatively long tail and a short sparrowlike bill—is a year-round resident from the brushy hillsides in Oregon, south through California and the Southwest into Mexico. Mexicans call the bird *la viejita*, meaning "little old woman." It is not shy around gardens and dooryards. The brown towhee probably mates for life, and may live 8 or 9 years.

A green-tailed towhee, which summers in the high mountains and plateau country, is a great rarity in my brother's garden. One appeared on a December day, took bread from his feeder, and scratched under the bushes for the birdseed mixture my brother had spread there. The green-tailed is the smallest of the towhees, 6 to 7 inches long, has a red-brown cap, white throat, and olive green back.

A rufous-crowned sparrow, usually a resident in parts of California south into Mexico and Baja California, came to my brother's garden and ate the baby-chick food in the company of two Brewer's sparrows. This bird is small, only 5 to 6 inches long, and resembles a chipping sparrow with a rufous crown and black-striped back. It is distinguished by the black "whisker" marks on either side of its throat.

## GRIT FOR SEED-EATING BIRDS

I remember, when I was a youngster, my father had an expression that used to puzzle me. Whenever oranges or some other fruit or vegetable weren't to be had in winter, he would say that they were "scarce as hens' teeth."

"But haven't hens got teeth?" I would ask him, and then we would both laugh at the foolishness of my question. Of course hens didn't have teeth! Anyone knew that if he'd just trouble himself to catch a chicken, open its mouth, and look at its smooth-edged bill.

That was long ago, before I knew much about birds. Since then I learned that some of them do have teeth, but they are in the stomachs of these birds, not in their mouths.

Some years ago I collected from sportsmen in upstate New York about fifty stomachs, or gizzards, of pheasants that they shot during the hunting season. That was when I was a government biologist and I wanted to get an idea of what kinds of fruits, seeds, and grains pheasants ate in autumn. If I knew, it would help me to tell farmers the kinds of trees, shrubs, and grains they should plant to attract pheasants. What I found did no more than prove what I already knew of their feeding habits from the studies of other biologists. But after cutting open fifty of their tough, muscular gizzards and looking inside of them, I knew, better than ever before, that grain- and seed-eating birds do have teeth.

Every one of those gizzards, each of them about the size of a flattened lemon, had masses of muscles overlying the thick, horny, ridged grinding surfaces of their inside walls. In almost every one I found, besides grains of whole corn and weed seeds, small pieces of limestone and gravel. These were the "teeth," or millstones, that ground up the corn and seeds under the pressure of the gizzard muscles and so helped the birds digest their food. I never forgot that experience and it helped explain a curious little drama that Ernest Harold Baynes, a noted bird-attractor of an older generation, told of many years ago.

Mr. Baynes, one cold snowy day, watched a flock of crossbills, bird-visitors from the Far North, swarm over a ruined building, where they spent hours nibbling at the mortar that held the bricks together. He got some of this mortar, pounded it up, and scattered it on the snow he had trampled down in his garden. Down came the crossbills to the mortar and they spent every day there for weeks eating it. The crossbills became so fearless that they permitted Mr. Baynes to walk about the flock while they fed. Some of them alighted on his hands and head, and even allowed him to pick them up. All the while they ate the mortar as if they were famished.

Mr. Baynes did not say why he thought these birds were so fond of this particular kind of grit, but mortar for bricks is usually made from calcium, or lime. There is supposed to be a mineral deficiency in the foods of northern finches and other birds, which, like the crossbills, usually live in the Far North. I believe that Mr. Baynes's flock of birds was just as starved for calcium as they were for the grit.

We would buy finely crushed oyster- or clamshells, which are high in calcium, and mix about 5 to 10 pounds of this grit with each 100-pound lot of our birdseed mixture. Coarse sand from the seashore makes a good grit for birds. The breeders of cage birds, particularly of the budgerigars, or "budgies," as they are popularly called, have discovered the value of suitable grit for these seed-eaters. They recommend grit not only for the budgies to grind their food and aid their digestion but to provide them with minerals, of which calcite, or lime, is most important. One noted breeder of seed-eating cage birds uses a grit containing 90 percent calcium carbonate (lime). He says that old mortar also makes good grit because it contains a lot of lime and certain helpful salts.

TRY THIS

For a calcium-rich grit supplement, mix 5 to 10 pounds of finely crushed oyster or clam shells into each 100-pound lot of birdseed mixture.

## SALT FOR BIRDS

Birds need salt for their health and well-being just as we do. Usually they won't eat table salt put out for them, but they will eat certain kinds of salty foods. Peanut butter and bread, both of which are favorite foods of birds, seem to have all the salt in them that a bird normally needs. A few years ago, I made two small compartments in our open feeder by nailing two strips of wood on the inside about 2 inches from each end. I keep the middle section of the open feeder filled with our seed mixture, but I fill these narrow end-compartments with peanut butter mixed with cornmeal and bacon fat. Chickadees prefer it to suet, and tree sparrows and juncos are fond of peanut butter in winter, especially during extremely cold weather. The peanut butter is high in protein and gives energy and warmth to birds that eat it.

## *"Marvel-Meal"— A Peanut Mixture for Birds*

RULE OF THUMB

Only offer birds peanut butter as one ingredient among others in a bird food mixture.

While living in Chapel Hill, North Carolina, I made one of the most rewarding discoveries of my years of experiments in attracting birds. I had tried mixing peanut butter with cornmeal because peanut butter alone sometimes sticks in the throats of small birds and causes their deaths. I remembered the experience reported in 1961 by the late Charles K. Nichols, research associate, Department of Ornithology, American Museum of Natural History, New York City. Nichols, an experienced bird-attractor, learned of the dangers of offering plain peanut butter to birds when he autopsied the bodies of a number of small, well-fed wild birds that had collapsed and had died at his feeders in New Jersey. He examined at least six of the birds and found the esophagus of each so crammed with the sticky peanut butter that he believed that it had caused them to choke to death. He recommended thereafter that this problem could be overcome by mixing the viscous peanut butter with other food ingredients, which would enable birds to eat it without danger of choking.

Some people I knew, with many years of experience in attracting birds and with whom I discussed this problem, had never known of a bird to die from choking on peanut butter. Some doubted it, and one skeptic suggested to me that the birds that died at Nichols's feeders may have succumbed from other causes. Of course this is possible, but to be sure that there is no potential risk for small birds that eat peanut butter, it is better, as Nichols suggested, to offer it to birds only in a mixture.

I soon discovered at Chapel Hill that cornmeal (ground corn, so much liked by birds) added to peanut butter still made an adhesive mess, but if I added some shortening and some flour, this would prevent the peanut butter from sticking to my hands. It also made a doughy mix of manageable consistency. When I offered the mixture to birds, the results were astonishing.

Within a few months during that summer of 1963 and in the following fall and winter, thirty-five species of North Carolina birds came to my yard to feed avidly on the mixture. I called it Marvel-Meal because of its marvelous attraction for birds and for its basic cornmeal so relished by them. The mixture also retained its soft consistency and ease of handling when I pressed it into the bark of trees in my yard or in cracks or compartments of my bird feeders. I also called the mixture the Bluebird Special because of its extraordinary attractiveness to bluebirds.

Marvel-Meal also had some advantages over other foods that attracted birds. I soon found that both the insect-eating and seed-eating kinds of birds seemed to have an insatiable appetite for it; therefore I needed this one food only to attract most kinds of garden birds. However, I continued to offer birdseed mixtures as supplemental foods. This helped provide a broad nutritional diet so necessary for birds.

RULE OF THUMB

Even if you have great success with Marvel-Meal, you should continue to offer supplemental birdseed mixtures to your birds.

I also found that the Marvel-Meal was far more attractive to birds than the suet I had been putting out for years for woodpeckers and other insect-eating birds. Another advantage was that the peanut butter mixture, besides protecting birds from choking, was cheaper than pure, or straight, peanut butter. And it can be offered to birds both in winter and summer with the assurance that it will attract more birds than either the suet or the seed mixtures.

Mammals seen eating Marvel-Meal at bird feeders in Chapel Hill include gray squirrels, flying squirrels, chipmunks, opossums, and raccoons. I will venture a guess that every species of bird in

## Marvel-Meal Original Recipe

The following mixture that I prepared periodically and stored in my refrigerator will make about 2 pounds of the doughy mix, which has the consistency of putty after it is thoroughly hand-mixed as one would mix a cookie dough.

1 cup peanut butter
1 cup Crisco or other shortening
4 cups cornmeal (white or yellow)
1 cup white flour

## Cheaper Marvel-Meal Alternative

1 cup peanut butter
2 cups melted beef suet
4 cups finely cracked corn
2 cups cornmeal (white or yellow)

First chop or grind suet into small pieces, put it into a pot, and melt it over a slow fire. When the suet has melted, stir in the peanut butter; remove from the stove, then stir in the other ingredients. Refrigerate until hard.

## Another Marvel-Meal Alternative

Another doughy, pastelike mixture that birds like can be prepared of 2 cups of bacon drippings (save in a tin can or other container) and 2 cups of cornmeal with flour added to bind the ingredients together. One winter day at Chapel Hill, Mrs. Matt L. Thompson was delighted when a ruby-crowned kinglet and then a pine warbler alighted on her hands as she was putting this mixture in one of her suspended feeding sticks (see illustration, page 26) in her garden. By experimenting with this and with other mixtures, you will discover the one that seems highly attractive to many kinds of birds and will be one that you feel you can afford to offer in quantity.

the United States or Canada that will come to feeders for seeds, fruit, or suet, if each samples Marvel-Meal, will come for it as avidly as the thirty-five species in my North Carolina yard.

Desirable as Marvel-Meal is to birds, its cost may make it prohibitive to many bird-attractors. In 1963, I made the mixture for about 20 cents a pound, but it would cost much more today. It is eaten so rapidly by birds that readers may wish a cheaper peanut butter mixture. I have prepared one, a harder mixture that birds eat less rapidly, have field-tested it, and know that many birds, including many or most of those that eat Marvel-Meal, are strongly attracted to it: for example, most woodpeckers that come to feeders, jays, crows, chickadees, titmice, and nuthatches.

At Chapel Hill, after preparing the mixture and before the suet had time to solidify, I poured or ladled it into small, round cardboard containers. I had already perforated two opposite sides of each container about halfway up from the bottom with a small hole. Then I had drawn a piece of white cord through each hole with a large knot at the end of the cord on the inside so that the cord could not pull out of the container. This gave me two loose ends or tying pieces with which I could fasten each container to the trunk of a tree or to a limb where the birds could cling to the bark and feed or perch on the edge of the cup to get at the contents.

WARNING

Cans with sharp edges can endanger pecking birds. It is better to offer birds food in cardboard containers.

The containers were about the size of those holding a half-pint of ice cream. These were of convenient size for handling and placing in the garden; however, larger cardboard containers (cans with sharp edges could be dangerous to the pecking birds) might be used with the bottom taken out before placement so that birds can feed from either open end.

After the mixture hardened, I put the filled containers in the refrigerator to cool and for storage. I discovered that this hard mixture slows consumption of the food by such rapidly eating birds as starlings, and it lasts much longer than the soft, puttylike mixture I call Marvel-Meal. However, I still prefer Marvel-Meal and recommend it as a superior attractant to birds.

# Water

Most of us, when we first begin to attract birds, forget that they need water, for both drinking and bathing. In some parts of the United States, particularly in Arizona, New Mexico, and other sections of the dry Southwest, water may be more difficult for birds to find than food. People who attract birds there have told me that all they ever put out is water, which draws birds to their gardens more readily than seeds and suet.

One winter day, a few years ago, I measured the approximate distance that a bird in our backyard would need to fly to get a drink of fresh water. It was two miles.

TRY THIS

A small birdbath can easily be installed on your window shelf feeder.

This meant that our birds, after eating the dry grain, birdseeds, salty peanut butter, and other foods that we offer them, had to fly four miles—two miles each way—to get a drink and then return to our yard. After that I never had to remind anyone in our family that our birdbath needed thawing at least once or twice a day when our out-of-doors thermometer registered below freezing.

Many years ago, shortly after I had put up my first window shelf feeder, I decided to keep a saucer full of water on it during the winter. It was a very cold day with the temperature below zero when I filled the saucer with boiling hot water and placed it for the first time on the feeding shelf just outside my living room window. I hadn't any more than closed the window when a chickadee alighted alongside the saucer. The little fellow shivered as I had often seen birds do in cold weather and stood there for a moment eying the hot water from which steam was rising in a small cloud. Then he stepped into it, shivered again (I thought with pleasure), and stood for some time without moving. He did not drink, but finally hopped out of the warm water, took a sunflower seed, and flew away. In a few minutes he came back and again stepped into the saucer of water, which was still very warm. For a while he stood

**Pigeons Keep Out!**

If you wish to keep pigeons out of the feeders, staple or nail several 1/4-inch-diameter brass or copper bars (they won't rust) horizontally across the open sides of the bird feeder. Place the bars so that the spaces between them are about 1 1/2 inches wide, which is enough to admit chickadees, juncos, nuthatches, and other small birds to the feeder. Do *not* use wire instead of bars; birds will get their heads entangled in the wires and may injure themselves. (See page 238.)

there as though enjoying the warmth of the water or the steam arising from it. Then out of the saucer he hopped, took another sunflower seed, and flew up into a tree to eat it.

Since that day we have seen many other birds stand in our saucer of hot water, just as that chickadee had done many years ago. Of course they drink the warm water too, but on cold days they remain in it for such a long time that we are convinced they like to warm their feet quite as much as they enjoy drinking.

If you don't want to walk out in the backyard to thaw your birdbath on cold, freezing days, you can easily set up a small birdbath on your window shelf feeder. There it will be quite simple for you to open the window and bring the small shallow pan or dish indoors, where you can thaw the ice in it and replenish it with hot water in the comfort of your home. Always put hot water in the birdbath in winter, or buy a birdbath heater, because you may find, as we did, that birds like their feet warmed.

RULE OF THUMB

Always be sure your birdbath is filled with hot water, whether you refill it frequently or install a birdbath heater.

# *Helping Game Birds in Winter*

If you live in the northern United States or in Canada where the winters are severe, with snow covering the ground for months at a time, besides songbirds, you may want to help some of the wild game birds to survive. In some parts of the northern states, up to 80 percent of the ring-necked pheasants may die in winter, especially where no grain crops are raised, and, occasionally, a severe winter has killed off most of the bobwhite quail in New England, especially in the northern part. Game birds are tough and resourceful. But when bitter winds pile great drifts along hedges and woodland borders, covering the weed patches and waste grain, followed by sleet storms that hard-crust the snow and coat every tree and berry-producing shrub with glittering ice, ground-dwelling quail and pheasants may freeze to death, unless they find a dependable source of food.

Early in my career, when I was a field biologist with the Soil Conservation Service, I gave many illustrated lectures about our soil, water, and wildlife conservation work. Winter speaking tours took me from Pennsylvania through upstate New York and into New England. Wherever I spoke, before farmers' groups, 4-H clubs, schools, garden clubs, Boy Scouts, and rod and gun clubs, I found an instant and enormous interest among people in the winter feeding program we had been teaching farmers and suburban dwellers in the North.

Within the farm woodlands or along their thicketed borders, we got farmers to leave a few corn shocks, opened, to give quail and pheasants ready access to grain the farmers scattered there. Others built permanent lean-to or roofed-over shelters of sturdy logs, covered with branches of pine, spruce, or other evergreens. Under the shelters they placed wooden hopper-type feeders. The roofed shelters screened the grain spread on the ground from falling snow or rain, and the self-supplying grain in the hopper feeders remained dry, and when refilled once a week, offered quail and pheasants a constant food supply. A hopper feeder about a foot square and three feet high held enough grain for a week or two, and the average feeding station needed several quarts of grain or considerably more each week, depending on the number of birds visiting the station and the severity of the weather. Farmers started their winter feeding programs in October to get the wild game birds accustomed to the location of the feeding station.

To help farmers maintain these feeding stations, Boy Scouts and their leaders from nearby towns and members of the Rod and Gun clubs traveled to them. Often they went there on snowshoes or on skis to sweep away any snow that blew under the shelters and to replenish with food the ground under them and the hopper feeders. Grain was often donated by kindhearted townspeople and local feed stores.

If you live anywhere that snowfall in winter is heavy and prolonged, you may want to establish one of these feeding shelters in an open corner of your property. You can be almost certain that it will be visited by game birds if your yard or garden borders open fields or a woodland and hedgerows, the places where quail and pheasants winter. Scatter whole corn, buckwheat, wheat, or birdseed mixtures and scratch grain, including the very important grit (see pages 33-35) under your feeding shelter, and you may attract regularly groups of ring-necked pheasants and quail, from fall until spring.

A farmer I knew in western Pennsylvania took great pride in a covey of fifteen bobwhites that came regularly each winter day to his yard to feed. They came at almost exactly 4 o'clock each afternoon. One day when I visited him, we stood and watched through a window as his bobwhites came, running over the snow with little rushes and bursts of speed, to his feeding shelter. They had traveled there under the protection of a hedge that extended from a hillside above the farm to the dooryard below. After feeding, the bobwhites turned and moved slowly back up the hill, staying under the hedge that not only gave them protection from the weather each day, but sheltered them at night as they roosted there on the ground.

Sometimes, roosting on the ground in winter poses a special hazard for quail. All one winter, Mr. Lindley Collins of Marshfield, Massachusetts, fed a dozen bobwhites in his yard. A snowstorm came, followed by a freezing rain. For ten days the quail did not come to feed. Mr. Collins thought the covey might be imprisoned under the snow, and having followed them to their roosting place, he knew where they slept. He went there at once and broke a hole in the crusted snow. The next day the covey came to Mr. Collins to be fed. Later, he discovered that the quail had come out from under the snow through the hole he had broken in the hard surface. Because he knew intimately this covey's habits, Mr. Collins had saved their lives.

## *Should You Feed Birds in Summer?*

The first winter that I started to attract birds I didn't feed them through the following summer. I was told that birds don't need our suet and seeds in warm weather because there are plenty of insects and fruits in our parks, fields, and gardens for them to feed on. From what I had already learned about birds, I knew this was true, and I decided that putting out food for them in summer would probably be wasteful and fail to attract birds.

In June of that year I spent two weeks in southern New Jersey with my parents. One day someone who knew of my interest in birds told me of a man a few miles away who attracted them. I called him on the telephone and to my delight he told me that he fed birds in summer as well as in winter. At his invitation I went to see him the next day. I shall never forget that visit and what it taught me about the fun one can have from feeding birds in summer. My host had only one large feeder in the center of his yard, but he had planted along his property lines several shade trees, thickets of shrubs, and dense clumps of honeysuckle and grapevines in which birds could hide their nests. Cardinals, catbirds, robins, thrashers, and many other birds sat in his treetops or shrubbery, whistling, calling, or flying from thicket to thicket. While we watched, a flaming red cardinal flew to the bird feeder, fed there for a few moments, then flew to a honeysuckle bush in a corner of the yard.

"Watch this," my host said quietly. He walked to the bird-feeding station, took some English walnut kernels from his pocket (pecan nuts would have worked equally well), and spread them on the tray of the open feeder.

Instantly a clamor of young bird voices came from the honeysuckle thicket. The male cardinal popped out of the bush, flew directly to the feeder, and picked up a walnut kernel almost from under my host's hands. Quickly the bird turned about and flew back to the honeysuckle bush, where I could see it poke the piece of walnut down the throat of one of several noisy young cardinals, not long out of the nest, that crowded about the parent bird. Still the young birds called, perhaps more loudly than before. The female cardinal now came to the feeder. She, too, picked up some of the walnut meats, flew back to the young birds, and fed them. Both parents now worked feverishly to feed the clamoring young ones until all of the walnut kernels on the feeding tray had been carried away. The young birds still called loudly, but when the man moved away from the feeder and we walked out of their sight behind the house, they were silent.

"Don't let anybody tell you that birds aren't smart!" my host said, smiling. "Shortly after those young cardinals left the nest, the parents, in addition to feeding them insects, gave them walnuts that I had been putting in the feed tray for the old birds. The young birds took a liking to them and soon learned that whenever they saw me at the feeder, that meant *walnuts*. So all they have to do now is to make a lot of noise and the parents fill them with walnuts to shut them up."

It takes only one experience like this to convince you that birds in the backyard are interesting at all times of the year, and that feeding them in summer can be just as entertaining as in winter.

**CARDINAL**

| | |
|---|---|
| Number of Eggs Laid in Clutch: | 3-4 |
| Days to Hatch: | 12-13 |
| Days Young in Nest: | 10-11 |
| Number of Broods Each Year: | 2 to 4 |
| Lifespan: | 4 to 14 years |

*Cardinals range throughout the eastern United States, west to the prairies, and also in the Southwest. Their bowl-shaped nests are made of grasses, twigs, and rootlets and are concealed 4 to 5 feet up in the forks of twigs and sometimes on branches, in vines, trees, and in bushes.*

### A Deadly Fungus

Birds that eat wet and moldy feed may become infected with *Aspergillus fumigatus*, a deadly fungus. Cowbirds, grackles, juncos, house sparrows, and other garden birds may become infected. The disease, aspergillosis, progresses through stages: at first afflicted birds gasp and wheeze, then they begin to mope or to sit about with their feathers fluffed. Among seed-eating birds, crowded at feeders or at feeding places on the ground, the sick ones may continue to feed and to stand feebly in the bird feeder while doing so. One of the last signs of the disease may be severe diarrhea, after which the individual birds afflicted may fall over and die. Healthy birds become infected when they eat at feeders or on ground contaminated with the sick birds' feces. There seems to be no cure for the infection, and the best way to prevent its spread is this: Stop feeding for at least a week or two or until the sick birds have died or moved elsewhere, then clean the feeders with Lysol, or another disinfectant, before putting them back into use again.

To prevent it, use only a clean birdseed mixture, free of dust and dirt, and do not allow too much of the seed or grain to accumulate in the feeders (or on the ground), especially if the feeders are open to rain and snow. About once a day, allow the birds to eat all the feed in the open feeder. Keep the feeder clean. Unless you have several feeders in operation, and rotate their use, clean, scrub, and disinfect the feeders at least once a week.

## FOODS THAT ATTRACT BIRDS IN SUMMER

In June downy woodpeckers brought their young ones, just out of the nest, to the chopped suet we kept in our open feeding tray. It was a delight to see the parent woodpeckers pick up pieces of suet and poke them down the throats of the young birds. Ovenbirds, a small warbler that nests in woodlands throughout most of the northeastern states, brought their youngsters from a nearby woods to feed them some of our suet. The pine warbler and other small birds liked suet in summer, and even wrens would sometimes eat it.

We have been feeding birds in summer for many years now and we have added very few things to the seeds, suet, and peanut butter that we offer birds in winter. We have discovered that our birds seem to like peanut butter on warm days just as well as on cold ones. Friends of ours put out oatmeal or rice with raisins in it for the summer birds; bread, cake, and biscuits are always attractive to them at any time of the year.

Birds like walnuts, both in winter and in summer, but nuts are expensive. We have gone to the woods in fall and collected black walnuts, which we crack open and lay on our feeding trays, where catbirds, sparrows, brown thrashers, and wrens will come to eat them, even during hot weather. Blue jays will eat roasted unshelled peanuts that we put on our feeding trays for them, but they seem to prefer these in winter. Many birds like peanut hearts, which you can buy in some of the ready-prepared birdseed mixtures sold by feed companies. In fall and spring, crushed pecans will attract many kinds of birds, including migrating warblers that may visit your backyard.

Of all the foods that will attract summer birds—robins, bluebirds, catbirds, thrashers, thrushes, mockingbirds, and others—fruits seem best. Sliced apples, raisins, soft cherries, bananas, and even oranges will bring these birds to your feeders. During the summer, we would often put out half an orange on our feeding shelf to attract a pair of golden Baltimore orioles. A pair of these birds attached their nest every year to one of the trailing outer branches of our neighbor's tall wild black cherry tree.

**TRY THIS**

Fruit is the all-around best food to offer birds in the summer.

## *Unusual Foods That Birds Will Eat*

One summer my mother fed a pair of catbirds steamed raisins. Each morning she put out a dish filled with them on the back porch feeding tray, and by noon the catbirds and robins had eaten every one of them. One day she also put out some leftover sour cream for them to see if they would eat it. When the catbirds discovered it they couldn't seem to gulp it down fast enough. Later, on a morning when my mother didn't have sour cream, she put out milk for them instead.

The catbirds came to the saucer of milk, tried to eat it, and then gave it up. There they sat on the feeding shelf, crying their displeasure and peering at my mother through the window glass of the kitchen door a few feet away.

A man I knew who taught me a lot about birds once told me that "birds are like children, easy to love and easy to spoil." The more you get acquainted with them and cater to their wants, the quicker you will discover the truth of this man's observation. Even so, you won't like them any the less for it. In fact, I think you will admire birds as we do, for their ability to learn quickly who their human friends are, and then to let us know what they like and what they don't like in this world.

### Eggshells for the Feeder

Blue jays, in the spring of the year, will eat the eggs of smaller birds, and probably those of larger birds, too. It is quite possible that the eggshells, as well as their contents, supply a mineral need of blue jays at this time of the year. We save the broken shells of hens' eggs, crush them, and mix them with grit in our bird feeders during the spring and early summer. Many birds come to eat both our eggshells and grit, particularly the blue jays, and the supply disappears quickly. We hope that this helps to keep blue jays from eating the eggs of some of the robins, catbirds, song sparrows, and other birds that nest in our backyard.

### A Well-Balanced Diet for Birds

We have a menu for our backyard birds that seldom varies. If you include some of the following in your feeders all the year round, as we do, birds will get most of the vitamins, proteins, fats, carbohydrates, and minerals that will enable them to survive the winter and will attract them to your feeders in summer. We feed our birds:

- yellow corn—crushed or cracked for smaller birds—to supply vitamin A and carbohydrates
- proso millet, sorghum, and wheat for carbohydrates and energy-producing vitamins
- peanuts and peanut butter for protein
- sunflower seeds for especially high-quality protein
- peanut butter, pecan meats, and other nuts for fats
- bread and peanut butter for salt
- grit and eggshells for calcium and phosphorus
- fruits for carbohydrates and vitamins

## THE BIRD THAT LIKED PIECRUST

One of the most interesting stories about a backyard wild bird that I ever heard was that of a tufted titmouse, a small gray bird related to the chickadee. This bird was in the habit of alighting on a man's hand for the piecrust he fed it each day. Now lots of birds like crisp piecrust if it isn't saturated with fruit juice, but this particular titmouse not only ate it himself, but fed it to his mate during the twelve days that she sat incubating her eggs. The female titmouse was too shy to come to the man's hand. Whenever he appeared in the yard and she wanted some of the piecrust, she came to a nearby shrub and called until her mate appeared. The male would fly directly to the man's hand, get some of the piecrust, and then carry it to his mate. Each time the male faithfully fed her before he ate some of it himself. Strangely, after the young birds hatched, the female titmouse no longer would eat piecrust, but her mate apparently wanted her to eat it again. He would fly to the man's hand, get some piecrust, and then follow her about, trying to feed it to her, but she was utterly indifferent to his offers. Perhaps during the incubating period she may have had a craving for a certain kind of food, and the piecrust satisfied her need.

I have no doubt that birds, like ourselves, have strong desires for certain foods at times when their bodies particularly need it. The more we hear of experiments with the feeding habits of birds, the more convinced we are that they will eat almost any kind of food, if it satisfies a need. A bird-attractor in North Carolina watched a pair of song sparrows feed their youngsters corn, wheat, rye bread, bread and milk, cottage cheese, mashed potatoes, sausage, hominy, and walnuts that he had put in his feeders. A pair of brown thrashers fed their youngsters all of these foods and canned salmon besides! Another bird-attractor found that catbirds liked soda crackers and bread and butter, blue jays ate the skins of baked potatoes, and some birds ate cheese when

**PLAIN TITMOUSE**

| | |
|---|---|
| Number of Eggs Laid in Clutch: | 3-9 |
| Days to Hatch: | 14-16 |
| Days Young in Nest: | 16-21 |
| Number of Broods Each Year: | 1, sometimes 2 |
| Lifespan: | 5 to 7 years |

*The plain titmouse ranges west of the Rockies from Oregon to Baja California. Its nest of mosses and grasses is lined with feathers and rabbit fur and is built in the natural cavities of trees and in old woodpecker holes. A plain titmouse pair will also dig its own nesting cavity but often prefers to nest in a bird box.*

other foods didn't seem to interest them. We are satisfied to stick to those foods that are more natural for birds to eat, which scientists have discovered will provide them with a well-balanced diet, both in summer and in winter.

## *To Have and to Hold*

A well-balanced diet will help keep your birds healthy and contented; but, when spring comes, some of them that have been coming to your feeders will be faced with a problem far more critical than that of finding food. The breeding cycle has started and those birds in your garden that mate early will already have chosen their mates and will be looking about for nesting places.

If you have thickets of shrubbery in your yard, you are fortunate, because in them the catbirds, robins, chipping sparrows, and other birds that return in the spring usually build their nests. But to ten or twelve kinds of birds in the eastern United States, including many that eat garden insects, your shrubbery is useless at nesting time. These birds—some of them the chickadees, titmice, nuthatches, and woodpeckers that came to your suet feeders all winter—*prefer to nest in holes in trees*. Now in these days of "clean gardening" some people seldom allow the trees in our gardens or those along our suburban streets to have dead limbs or dead trunks in which these birds can, with their sharp beaks, excavate a nesting hole. Gardeners either cut off those dead tree branches, or seal the hollow tree cavities with cement. This means that the hole-nesting birds must seek a nesting place elsewhere, unless you supply them with homes. By providing shelter, as well as food, most of us can keep some of these birds about our gardens all summer.

# *Birdhouses in Your Garden*

In the small southern New Jersey village where I spent my boyhood, I remember a male flicker that, one spring, appeared to go crazy. These big woodpeckers nest in holes that they usually drill in dead trees. At that time, I didn't know why he sometimes drummed with his beak on resonant, hollow limbs, rooftops, or any other sounding board that he could find. Later, I learned that his long, rolling tattoo challenged other males, attracted the attention of his lady love, and played some mysterious part in the flicker's courtship and mating.

Early one morning, this particular male flicker rapped a few times with his bill on one of our neighbor's metal downspouts that carried rainwater from the roof of his house to the ground. The flicker must have liked the sound of his beak striking metal. Suddenly he attacked the downspout furiously until his thunderous rolling call awakened echoes from the houses all along our quiet village street.

*B-r-r-r-r-r-r-r-r-r-r-r-r-t-t-t-t-t-t-t-t!* The sharp tattoo rang out like the rattling of a snare drum. People came out of their houses curiously and then laughed when they found that the noise came from a lovesick woodpecker.

After several days of this racket, our neighbor didn't laugh. The flicker had drummed a hole right through his galvanized rainspout. Then someone said that the bird hadn't been drumming after all, that he had been drilling a nesting hole for his mate. Apparently the flicker discovered his mistake, because he flew away and no one heard him drumming on the downspout after that.

There is both comedy and pathos in this story. The flicker is a wonderfully useful bird about our gardens, where he destroys colonies of ants in our lawns, and insects that eat our garden crops. Like other woodpeckers, this bird has been deprived of nesting homes wherever people cut down dead trees or trim off the dead branches of living trees. Perhaps that flicker of my boyhood had attacked the metal downspout because there were no hollow limbs about in which he might have carved a home. It might explain another story about a flicker that I once heard.

A farmer in southern New Jersey built a new barn, which he completed in the fall of the year. The following spring, a flicker came to the empty barn and drilled so many holes in its sides that it looked as if it had been cannonaded. I am sorry that I wasn't there to see the look on this foolish bird's face when he finished drilling each hole and poked his head inside the barn. What a big room he had cut into! He must have had great faith to keep on with his drilling.

In a way, I am glad that I *wasn't* there when the farmer saw those holes in his new barn. I might have had a lot of trouble convincing him that the flicker didn't know his barn from a dead tree. It might have been even more difficult to make him believe that he, himself, had invited the flicker to his barn by cutting down all the dead trees in his woods in which this poor bird might have built its home.

## *The Adaptability of Hole-Nesting Birds*

One of the wonderful things about the flicker is his adaptability. He and more than thirty other kinds of hole-nesting birds in the United States have learned to accept a man-made home—the bird nesting box. He will now raise his family in one of these in place of the holes that he ordinarily digs for himself in the rotted limbs and the trunks of dead, or partly dead, trees. Perhaps if my boyhood neighbor had known this, and had put up a flicker nesting box in his yard, he might have prevented that flicker from damaging his rainspout.

Flickers are not usually difficult to attract. Of the nine kinds of woodpeckers that breed in our northeastern states, the flicker seems most likely to nest in our city parks and suburban gardens.

**COMMON FLICKER**

| | |
|---|---|
| Number of Eggs Laid in Clutch: | 5-10 |
| Days to Hatch: | 11-12 |
| Days Young in Nest: | 25-28 |
| Number of Broods Each Year: | 1 or 2 |
| Lifespan: | 5 to 13 years |

*There are three varieties of the common flicker. The yellow-shafted flicker of the eastern United States nests from Canada, east of the Rocky Mountains, east to the Atlantic coast, and south to Texas and the Gulf coast of Florida. The red-shafted flicker, which is about the same size as the yellow-shafted, nests westward from the Great Plains and is a year-round resident of the Rocky Mountains down to the Pacific coast and south to Baja California. The gilded flicker, a small desert bird of south Arizona and southeast California, usually digs its nesting and roosting holes in the giant saguaro cactus.*

Nesting boxes for them, unless you choose to buy one from a dealer in bird-attracting equipment, must be built according to a specified size or the birds won't nest in them. The circular entrance hole must be large enough for the bird to enter and to leave the nesting box easily, yet not so big that a larger bird can get in and drive the flicker out.

**RULE OF THUMB**

The entrance hole to a nesting box should be the exact size required for whatever bird it is built for and no larger.

I know of a flicker nesting box in which the builder cut the entrance hole so large that it would admit a screech owl. Now the screech owl is a beautiful and interesting bird that has nested in my own backyard. We like screech owls and welcome them, but sometimes they will kill and eat flickers. That is why you should measure the dimensions exactly if you build the flicker nesting box described on page 57.

The man who made that extra-large hole to his flicker box was fortunate. A screech owl that had been dispossessed from her own home in a hollow tree nearby decided to adopt the young flickers, instead of choosing to eat them as she might easily have done. Frequently this frustrated mother owl came to the flicker box to squat like a setting hen over the young flickers. Once she even brought them part of a mouse, perhaps intending to feed it to them. All this time the parent flickers were feeding their young ones ants and other insects, which a normal young flicker requires. Eventually, the screech owl left the nest box, leaving the baby flickers unharmed.

A scientist interested in bird behavior might have explained this screech owl's unusual actions. I think he would have said that her instinctive desire to feed and mother young birds, even though they weren't her own, was stronger than her normal hunting instinct. Ordinarily, screech owls feed on mice, beetles, moths, and other creatures of the night, but occasionally they will kill and eat small birds. When I heard this astonishing story it taught me a simple lesson: Always make the entrance hole to a flicker nesting box, or to any other birdhouse, the exact size required for the particular bird and no larger.

## The Design of a Bird Nesting Box

This table is from *Homes for Birds,* a free government pamphlet. See appendix 3 for information on how to get it.

| Kind of Bird | A: Size of Floor (inches) | B: Depth of Bird Box (inches) | C: Height of Entrance Above Floor (inches) | D: Diameter of Entrance Hole (inches) | Height to Fasten Above Ground (feet) |
|---|---|---|---|---|---|
| Bluebird | 5 x 5 | 8 | 6 | 1½ | 5-10 |
| Chickadee | 4 x 4 | 8-10 | 6-8 | 1⅛ | 6-15 |
| Titmouse | 4 x 4 | 8-10 | 6-8 | 1¼ | 6-15 |
| Nuthatch | 4 x 4 | 8-10 | 6-8 | 1¼ | 12-20 |
| House wren and Bewick's wren | 4 x 4 | 6-8 | 4-6 | 1¼ | 6-10 |
| Carolina wren | 4 x 4 | 6-8 | 4-6 | 1½ | 6-10 |
| Violet-green swallow and tree swallow | 5 x 5 | 6 | 1-5 | 1½ | 10-15 |
| Purple martin | 6 x 6 | 6 | 1 | 2½ | 15-20 |
| House finch | 6 x 6 | 6 | 4 | 1½ | 8-12 |
| Starling | 6 x 6 | 16-18 | 14-16 | 2 | 10-25 |
| Crested flycatcher | 6 x 6 | 8-10 | 6-8 | 2 | 8-20 |
| Flicker | 7 x 7 | 16-18 | 14-16 | 2½ | 6-20 |
| Golden-fronted woodpecker and red-headed woodpecker | 6 x 6 | 12-15 | 9-12 | 2 | 12-20 |
| Downy woodpecker | 4 x 4 | 8-10 | 6-8 | 1¼ | 6-20 |
| Hairy woodpecker | 6 x 6 | 12-15 | 9-12 | 1½ | 12-20 |
| Screech owl | 8 x 8 | 12-15 | 9-12 | 3 | 10-30 |
| Saw-whet owl | 6 x 6 | 10-12 | 8-10 | 2½ | 12-20 |
| Barn owl | 10 x 18 | 15-18 | 4 | 6 | 12-18 |
| American kestrel | 8 x 8 | 12-15 | 9-12 | 3 | 10-30 |
| Wood duck | 10 x 18 | 10-24 | 12-16 | 4 | 10-20 |

The simple design of the birdhouse shown here will suit *all* birds that will nest in bird boxes, but the *size* of your birdhouse and its dimensions must be made to suit each kind of bird as shown in the table above. If you want to attract bluebirds, you must build a bluebird house. If you want to attract house wrens, you must build a house wren box. Make the roof of the house or one of its sides detachable or hinged to allow you to inspect the inside or to clean it (see page 70).

# Where to Place Nesting Boxes

The first year that I put up a flicker house, I lived in the country. Several large apple and pear trees grew in the backyard and it would have been easy to nail the flicker box to the trunk of one of them. Had I done so, however, the nesting box would have been too heavily shaded by the leafy boughs. Most birds like their nest boxes to be in full sun or at least in a place where they will get sunshine most of the day.

RULE OF THUMB

The best place to mount a nest box is in the full sun or in a spot where it will get sunshine most of the day.

Hole-nesting birds are also wary sometimes of nesting in a bird box on the main trunk of a tree. Perhaps they have learned from experience that cats and raccoons can climb to such bird boxes and kill the young ones, and the parents too, if they catch them in the nest box. I might have nailed a 30-inch-long sleeve of tin or sheet metal tight around the tree trunk to keep cats from sinking their claws in the bark and, thus, from climbing the tree, but the metal guard could have injured or killed the growing tree by cutting into the bark.

A post seemed best that first year, and so I cut one about 12 feet long and 6 inches in diameter from the trunk of a cedar tree that I'd had to cut down because it grew too close to my house. After I had fastened the flicker nesting box very firmly to the side of the post, near its top, I dug a hole 3 feet deep near the edge of my lawn at the rear of the yard and set the post in it. Before I filled the earth in around the post and tamped it solidly, I turned the post until the entrance hole of the nest box pointed toward the southeast, which was *away* from the prevailing winds. This would keep rain from driving into the birdhouse in bad weather.

Here, in the open, my flicker box had plenty of sun, it was 9 feet up in the air and above the surrounding shrubbery. Flickers like to nest in boxes that are from 6 to 20 feet above the ground. At 9 feet, it was high enough to satisfy the flickers, and I could still

## The Flicker Nesting Box

Make your flicker nesting box 7 inches square (inside dimensions) and 16 to 18 inches high, from bottom to top. Bore the 2½-inch-diameter entrance hole in the front panel about 14 to 16 inches above the floor of the box and drill several ¼-inch holes just above the entrance hole for ventilation. Drill several l-inch holes in the bottom of the box to drain away any rainwater that may get inside. Complete the bird box by putting a roof on it. The roof should slant toward the front of the box to shed rainwater. It should also project from the front of the box several inches, like a porch roof, to shelter the entrance hole in the front panel from wind-driven rain. Be sure to fasten the roof or one of the side panels with hinges so the box may be easily cleaned out once each year.

After the box is assembled, select a board about 1 inch thick, 12 inches wide, and 24 inches long. With screws, fasten it to the back of the birdhouse, with its 24-inch length pointing from sky to ground. This "mounting board" will be useful to nail, or preferably to screw, the assembled birdhouse to a tree or the side of a post.

Last—and this is most important—put a 2-inch-deep layer of moist earth and sawdust, or moist earth and wood shavings, on the inside floor of the box. Apparently all woodpeckers need this material to hollow out a bed in which to lay their eggs. Ordinarily, when they dig a vertical hole in a dead limb for their home, they use chips and rotted wood from their excavations to line the bottom of the nesting hole. Unless you put a layer of sawdust, wood, or cork chips in the bottom of a flicker box, these birds may not nest in it.

RULE OF THUMB

To keep rain from entering your nest box, point the entrance hole away from the prevailing winds.

stand on a 6-foot stepladder and reach the box to inspect it occasionally, or to clean it once a year.

Last of all, I wrapped a metal sleeve, 30 inches wide, around the nest-box post to keep cats from climbing it. I nailed the cat guard snugly in place, then I sat back to wait for my first pair of nesting flickers.

## WHEN SHOULD YOU PUT UP BIRD NESTING BOXES?

My notes say that it was about May 1 when I put up that first flicker nesting box many years ago. Flickers usually begin to nest in New York State about the middle of May. Although I didn't realize it at the time, it is best to have all bird boxes in place in late summer, before the leaves fall from the shrubs and trees. This will prevent you from putting them in places that are too densely shaded, especially if you put up the boxes near or under trees or shrubbery. Setting them up before fall will also permit them to weather over winter, and probably be more acceptable to birds in the following spring. More important, wintering flickers, nuthatches, downy woodpeckers, chickadees, and other hole-nesting birds will have a

**RULE OF THUMB**

To ensure that you don't mount your nest boxes in places that are too densely shaded, have all your bird boxes in place by late summer, before the leaves have fallen.

### Bluebirds, Chickadees, and Other Small Birds

Bluebirds, chickadees, and other small birds that have nested in the abandoned nest holes of woodpeckers (A), probably for thousands of years, will accept the man-made nesting box (B), if it is designed correctly. The man-made nesting box should be built to approximate the design originally carved out of a tree by a woodpecker.

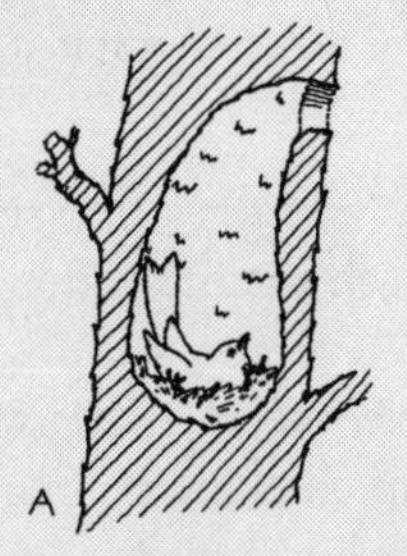

sheltered place to roost on cold, stormy nights. If you can't get your birdhouses up by the fall, you can put them up anytime during the winter. Your chances of getting birds to nest in them will be improved if you have them in place before March 1.

That first flicker house I put up was an exciting adventure. A pair of flickers came to it one day and inspected it carefully. The male bird, which I knew from the female by his black facial markings, came first; and, after a great deal of going in and out of the box, the female joined him. I don't know when she finally accepted this home of his choice, but from that day on they were in and out of the box regularly.

I felt a lot of pleasure in knowing that the birds had accepted my handiwork. Yet, knowing what I know now about birds, I shouldn't have been surprised if I hadn't got a pair of flickers nesting there the first year. I have learned that birds, during the first nesting season that a box is up, may not use it. Sometimes they may for several years ignore a nesting box that meets every requirement of size, height above the ground, apparent safety, and required position in full sunshine or in part shade.

## THE FIGHT BETWEEN THE FLICKERS AND THE STARLINGS

Birds of different kinds, like the flicker and the starling, do not usually fight when they are feeding near one another on your lawn. You will see a flicker stand in one spot and probe deep with its bill into an anthill to eat these insects. While he does so, a starling a few feet away may be strutting about and stopping occasionally to thrust his sharp beak deep in the grass. The starling is probably digging out the white Japanese beetle grubs that move up near the surface of the ground in spring. Apparently there is little, if any, competition between the flicker and the starling for insect foods or for territories in which to live. But when they go apartment hunting, especially when "bird apartments" are scarce, a lot of things can happen.

One day trouble came to the flickers in my nesting box. I heard them screaming and, when I ran out into the yard, I saw the male flicker pulling at the black wings of a starling that had gotten inside the flickers' nesting box. Starlings nest in holes in trees and in bird boxes, too. They are smaller than flickers and can easily get into flickers' nesting holes and nest boxes. The fight for the house was furious. The flicker pulled the starling out of the hole and wrestled it to the ground, but while they were battling, the mate of the fighting starling slipped into the bird box and held it by pecking sharply at the female flicker whenever she tried to get in the entrance hole.

### Building Materials for Birdhouses

If you build your own birdhouses, a fascinating indoor project for winter evenings, you will want your houses sufficiently well made to last five years or longer. Use rough, unplaned wood 3/4 inch to 1 inch thick, in preference to thinner wood, metal, clay, or building paper. Wood is the best insulator against heat and cold; a wooden bird box will generally be cooler in summer and warmer in winter than one made of metal or pottery. Soft woods—cypress, white pine, cedar, yellow poplar, and whitewood—are easy to "work," but many conservationists now avoid tropical and exotic woods so as not to encourage depletion of tropical rainforest bird habitats. You should also avoid using pretreated woods, which contain arsenic and might harm nesting birds. Cypress birdhouses will be the most durable of all. Do not use green or wet lumber because it will split and warp after it dries. We always use well-seasoned wood to build our bird nesting boxes.

For long-lasting houses, you should put the parts together with brass screws, brass hinges, and galvanized or brass nails (steel nails will rust). Bird boxes put together with screws are always more solid and durable than those that are nailed. Sharp changes in temperature will cause nails to loosen and pull out of the wood.

The starlings and flickers, both male and female, fought for a long while, but I did not interfere. I have learned it is a hopeless struggle to try to keep starlings away, and I have never discriminated against any of the birds in our yard. All are welcome to our feeders, and the birdhouses belong to the birds that want them and can hold them. Later that day the starlings were in full possession of the nest box and the flickers had disappeared.

I was angry at the starlings because I wanted flickers in my yard, but starlings have the useful habit of feeding themselves and their youngsters on army worms, potato beetles, white grubs of the May beetle, billbugs, and other insects that eat our garden crops. I thought about the problem of getting the flickers to come back and finally came up with what I believed was an answer. A few days later I built another flicker box and put it on a post in another part of the yard. I was confident that the new box would soon have tenants because of the bird housing shortage, but that spring and summer passed and no flickers came to nest.

The following year, flickers, perhaps the same pair, came back, not to the newest nesting box, but to the one that the starlings had dispossessed them of the previous spring. I had cleaned it out after the starlings had finished their nesting and the flickers were soon quite at home. When a pair of starlings decided to nest in our yard, they pleased me by going to the new nesting box. As far as I knew, they had no quarrels with the flickers that summer and both pairs successfully brought up their broods of young.

## PAINTING OR STAINING YOUR BIRDHOUSES

Many people paint or stain birdhouses after they build them, or treat each part of the house with a wood preservative before they put the bird box together. Recent research seems to show, however, that painting, staining, or preservative chemicals aren't necessary, and can keep birds from using the house longer than once believed. A shingle can be a simple alternate way of waterproofing the roof of your birdhouse.

In *all* bird boxes, I always drill several 1/4-inch holes in the front panel, *between the roof and the entrance hole*. These will ventilate the box and keep the young birds cool on hot days. Do *not* drill air holes below the entrance hole. These will create a draught on the young birds in the nest and might endanger their lives. Besides ventilation holes, I also drill 3/4-inch holes in the floor of all nest boxes, just as I did in the flicker nest box. These will drain off any rainwater that may get in.

Do *not* put a perch on the outside of a birdhouse. Wrens, woodpeckers, chickadees, and other native birds that nest in man-made birdhouses do not need it. The perch will give English sparrows and starlings a place to cling from which they can harass the rightful occupants, and possibly evict them.

## *Other Hole-Nesting Birds You May Attract*

If you live in the eastern states and aren't too far from a park or a woodland with big trees, it is possible for you to attract downy and hairy woodpeckers, chickadees, nuthatches, and tufted titmice in addition to flickers. In many regions, these are resident birds that do not migrate but live in your area all the year round.

TRY THIS

It's a good idea to continue feeding birds through spring and summer because it will encourage them to raise their young in the trees, shrubs, or nest boxes in your yard.

Some of them may be the ones that came all winter to your feeders for peanut butter and sunflower seeds. If you continue your feeding of these birds through spring and summer, it may encourage them to stay in your yard and raise their families in bird boxes, instead of going to the woods to hunt for a natural cavity, or to drill a hole in a tree for themselves.

One spring a pair of downy woodpeckers nested in a bird box I had put up in an apple tree in my yard to attract nuthatches. The male of this pair had a deformed bill. It curved upward as if he had broken it and by this sign I knew him. He, and his mate (which I

### Chickadee, Titmouse, Nuthatch, and Downy Woodpecker Boxes

For the smaller hole-nesting birds I build a simple birdhouse that is similar in design to the flicker nesting box, but is not nearly so large. The entrance hole of 1¼ inches in diameter, the inside dimensions of 4 inches by 4 inches, and the height of 8 to 10 inches from the floor of the box to the inside bottom of the roof will accommodate titmice, nuthatches, and downy woodpeckers. The entrance hole of this box should be 6 to 8 inches above the floor of the box. A box built to these specifications will also suit chickadees if you make the entrance hole 1⅛ inches in diameter, in stead of 1¼ inches.

assumed to be the same bird), nested in this bird box for two consecutive years. Bird scientists have proved by trapping and banding birds that pairs of downy woodpeckers and pairs of black-capped chickadees often remain mated and live together continuously for two or three years. Our male downy woodpecker with the crooked bill disappeared after the second year, as many small birds do. The death rate among them is high.

## NESTING BOXES FOR SMALLER BIRDS

Some birdhouses that you can make, or buy if you wish, will suit several kinds of the small birds that nest in holes in trees. If you prefer to buy your bird boxes, always tell your dealer the kind of a nest box you want. For example, ask for a *chickadee* nest box, a *house wren* box, or a *flicker* box, because each one is made in different sizes and of certain specifications to suit each kind of bird. See the dimension table and typical drawings throughout this chapter. The kinds of birds that you will be able to attract to your nest boxes will depend a great deal upon where you live. If you live near a park or woodland where large trees grow, you may be able to have chickadees, nuthatches, and titmice living in your backyard. If you

**TUFTED TITMOUSE**

| | |
|---|---|
| Number of Eggs Laid in Clutch: | 5-6 |
| Days to Hatch: | 13-14 |
| Days Young Are in Nest: | 17-18 |
| Number of Broods Each Year: | 1, sometimes 2 |
| Lifespan: | 3 to 12 years |

*The tufted titmouse ranges throughout the eastern and southern United States and is common mainly east of the Great Plains. Its nest is made of wool, moss, hair, cotton, and leaves and is built 3 to 90 feet from the ground in the cavities of trees, in old woodpecker holes, in hollow metal pipes, in holes in old fence posts, and in birdhouses.*

### How to Put Up a Bird Nesting Box

*Wrong way at A.* If box is to be attached to a tree, do not put it up with the entrance hole facing upwards, which would allow rain to enter the box. *Correct way at B.* Attach box to the tree so that the entrance hole faces slightly downward. When attaching the bird nesting box to a post *(C)*, put it either on the top or on the side of the post. Be sure to wrap a metal sleeve around the post below the nest box to prevent cats, raccoons, and squirrels from climbing to it.

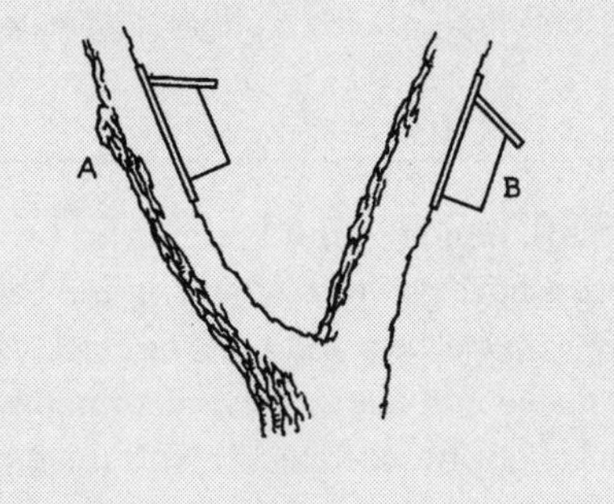


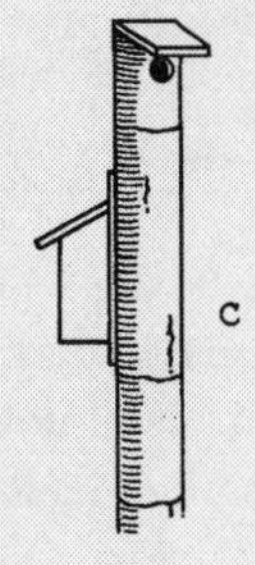


RULE OF THUMB

Be sure your dealer knows what kind of bird you want to attract before you buy a nest box.

live out in the suburbs or country where open fields surround your property, you may be able to attract bluebirds, tree swallows, and purple martins to your nest boxes.

Chickadees, titmice, downy woodpeckers, and nuthatches are birds of the woodland. They seem to prefer rustic nesting boxes made of slabs of wood with the bark on them, which resemble the outside of the cavities in trees where these birds usually nest. They also will nest in birdhouses made of weathered lumber. If your yard is close to a wooded park or an old orchard, you may attract these birds. Chickadees and titmice will nest from 6 feet above the ground to 15 feet or higher; downy woodpeckers, 6 to 20 feet above the ground; and nuthatches, 12 to 20 feet. These are among the few birds that may be attracted to a nesting box put up in a shaded place, because they ordinarily nest in woodlands and woods borders.

Place each box on a post (which is preferable to a tree), and put a metal sleeve around each post to keep cats and raccoons from climbing it. It is a wise precaution to set the post and bird box at least 8 feet away from a fence or tree from which a cat might jump to the nest box. When we put up a nesting box in a tree or on a post, we face the entrance hole away from the direction of the prevailing summer winds.

RULE OF THUMB

Remember to equip your post with a metal sleeve and to set the post in the ground at least 8 feet from any fence or tree from which a predator might jump.

## HOW MANY BIRDHOUSES SHOULD YOU PUT UP?

Some of our hole-nesting birds will live surprisingly close together, and peaceably, *if they are of different kinds*. I remember, years ago, a telephone pole in Massachusetts that was an apartment dwelling for a pair of flickers, a pair of red-headed woodpeckers, and a pair of chickadees. The families of these three different kinds of birds lived in separate "flats" or holes that were only a few feet above or below each other. Each pair of birds hatched and brought up its young without a single quarrel as far as anyone knew.

Now if all three pairs of these birds had been of one kind—say, flickers—they might have spent most of their time fighting over their territorial rights instead of raising their families. As a general rule (except swallows and certain other birds that nest in colonies) birds of the same kind will not usually nest close together. This means that in your small yard in the suburbs, or in the country, it would be best to put up no more than one birdhouse for each kind of bird you want to attract.

RULE OF THUMB

Since birds of the same species will generally not nest close together, there is usually no point in putting up more than one birdhouse for each kind of bird you want to attract.

For example, in our suburban New York City backyard, we had flickers, downy woodpeckers, starlings, house sparrows, and house wrens nesting in our bird boxes during the same year. A woodland was nearby and it was theoretically possible to have the woodland hole-nesting birds—crested flycatchers, hairy woodpeckers, chickadees, titmice, and nuthatches—also nesting in our yard. Like the fisherman who baits more than one hook, we put

RULE OF THUMB

You can mount more than one birdhouse in the same tree, each at a different height, but it's a good idea to space them at least 25 feet apart.

up a box for each of these kinds of birds, hoping to attract them.

Some of our bird nesting boxes, for different kinds of birds, were in the same trees, but at different heights above the ground. We made it a general rule to space them no closer together than about 25 feet. Although birds of different kinds will sometimes live peaceably much nearer to one another, we like to keep our boxes as far apart as a neat arrangement of them on posts or trees in our yard will allow.

## CRESTED FLYCATCHERS AND HAIRY WOODPECKERS

More than seventy years ago when I first began to notice birds, the one that seemed strangest to me was the great crested flycatcher. Each spring, in late April or early May, when I heard its loud cry of *w-h-e-e-e-e-e-p!* ring through the southern New Jersey woodlands, I knew that it had returned to spend the summer.

**Crested Flycatcher Nest Box**

Make the entrance hole of the crested flycatcher nest box 2 inches in diameter, the inside dimensions of the box 6 inches square and 10 inches deep from floor to inside top of box. Cut the entrance hole about 8 inches above the floor of the box. Crested flycatchers like their nesting boxes set on a post or tree between 8 and 20 feet above the ground.

The crested flycatcher had one habit that puzzled me and lent a deeply mysterious air to this hole-nesting bird of the woods. In most of its nests I found almost invariably a piece of dried snakeskin. Usually there were only a few pieces in each nest; sometimes an entire cast-off snakeskin lay draped over the outer edge and trailed outside of the cavity in the tree in which the nest had been built. Some of the older and wiser people of my home village had a solution to this problem. They said that crested flycatchers put snakeskins in their nests to frighten away any animal that might try to eat their eggs or young ones.

Long before I got interested in crested flycatchers, naturalists had puzzled over this strange habit and found a much more logical explanation for it. One man in West Virginia, who studied the ways of crested flycatchers, identified the cast-off skins of at least five kinds of harmless snakes that these birds had woven into their nests. Yet, in some nests, crested flycatchers hadn't used the moulted skins of snakes at all. Instead they had added pieces of the shiny outer skins of onions, waxed paper, paraffin paper, strips of cellophane, and other materials that resemble dried, cast-off snakeskin. Apparently the shininess of snakeskin makes it attractive nest material to crested flycatchers, not any supposed protective value it gives its nest.

Crested flycatchers will nest in bird boxes that are not too far away from woodlands. The same pair came to a nesting box in a Pennsylvania backyard for three consecutive years and one of the pair lived to be at least eight years old.

## NESTING BOXES FOR WRENS

Of all hole-nesting birds, wrens are the ones you are most likely to attract to your city, suburban, or country backyard. Several kinds will nest in bird boxes. There is hardly a section of the United States where one of the hole-nesting species isn't found. These tiny, explosively energetic birds bubble with song and nervous excitement. If you put up a nesting box for them and get a pair of perky little house wrens, Bewick's wrens, or Carolina wrens nesting in your garden, they will entertain you endlessly. The Carolina wrens and Bewick's wrens are resident birds and will usually remain with you all the year round. The house wren is migratory. Each fall the little house wren leaves the northern states and spends the winter in Louisiana, Georgia, Florida, and other southern states. It returns north again in the spring.

Before our country was settled, the house wren nested in a cavity in a tree or stump, usually in natural openings or burned-over lands in the wilderness. Today, these little birds are the most eccentric in their choices of nesting places of any that I know.

One spring, at a bird sanctuary on Wallops Island, Virginia, the caretaker picked up twenty-four bleached cow skulls that he found lying about on the island. At various places he hung them up in trees and shrubbery. Almost immediately house wrens moved into twenty-three of the skulls, and reared their families in them.

They have built their nests inside the abandoned paper nests of

### The Hairy Woodpecker Nest Box

The hairy woodpecker, a highly useful woodland bird, like its smaller cousin the downy woodpecker, will also accept a bird box with an entrance hole 1½ or 2 inches in diameter. The box should be 6 inches square (inside dimensions) and about 14 inches deep from floor to inside top of box. Cut the entrance hole in the front panel about 11 inches above the floor. Put your nesting box for the hairy woodpecker on a pole or in a tree between 12 and 20 feet above the ground.

### Boxes for House Wrens, Bewick's Wrens, and Carolina Wrens

For house wrens and Bewick's wrens, make the entrance hole a slot, instead of a round hole. We make ours from 1 to 1¼ inches high and about 3 inches long. This slot, which looks like a letter drop in a city mailbox, gives the wrens more room to maneuver the long twigs that they carry into the box as a foundation for their nests. For the slightly larger Carolina wren, cut the entrance slot 1½ inches high and 3 inches wide. For all three kinds of wrens, make the inside of the house 4 inches square and about 7 inches high, measured from the floor of the box to the inside bottom of the roof. Cut the entrance slot in the front panel of the wren boxes from 4 to 6 inches above the floor. See the dimension table and typical drawings early in the chapter.

the white-faced hornet, in the deserted nests of barn swallows, robins, and orioles, in a fishing creel hung on the side of a fence, in rusty tin cans on garbage piles, in the pocket of a pair of overalls hanging from a clothesline, in an iron pipe railing, and in many other unusual places.

House wrens will nest in a birdhouse suspended in the air from a tree, but in our experience, all other birds that nest in birdhouses prefer them set firmly on a post or attached to some other solid object.

TRY THIS

Wrens can be an exception to the rule that you should put up only one nesting box for each kind of bird.

With the wrens we make an exception about putting up only *one* nesting box for each species of bird. I have hung three in my backyard because of the tremendous nest-building urge of the male bird. The males usually arrive in the North several days ahead of the females and at once start to build their nests.

One spring a male built three nests in our bird boxes and then built another in our neighbor's wren nesting box, all before the female arrived. The following day the female followed him about to each nesting box that he showed her, between bursts of singing.

When she finally decided upon one that seemed to please her, she immediately began to throw out every stick that he had worked so hard to put into the box. We thought the male wren would explode with anger, but later we saw him helping the female build a new nest in the box of her choice.

Occasionally the industry of the bustling house wren catches him an extra wife. While his mate is incubating her first set of five to eight eggs, the male, between bursts of singing, may carry sticks into a nearby nest box. There his singing and nest-building sometimes attract another female. Friends of ours once thought they had two pairs of wrens in the two boxes in their backyard. Then they watched and discovered that only one male was helping the two females feed the youngsters that filled each nest box.

One of the most pathetic bird stories I have heard was that of an unmated female house wren in Maine. This little bird, by herself, built a nest in a wren nesting box. After that, she allowed no other bird to alight on her house or to come near the nest. If they did, she was ready to fight them, no matter how large they were. She remained at the nesting box all that summer until August, when she disappeared. During that time, the housewife who watched the wren never saw a male house wren about, nor heard one sing. After the female left, the woman took the box down. Inside it she found an exquisitely built nest containing twelve eggs, which were all sterile.

Another wren, the Bewick's wren, a gentle and confiding creature, is also a familiar bird in gardens. It is a little larger than the house wren and has a white stripe over each eye. Its range overlaps that of the house wren in parts of our northeastern states, except in New York State and New England, where this bird is practically unknown. This little wren and its close relatives are widely distributed over the United States from central Pennsylvania and Virginia to the Pacific coast. The Bewick's wren has a sweet, tender song, considered one of the finest of all birdsongs. Many people would rather have the Bewick's wren about than the house wren,

| HOUSE WREN | |
|---|---|
| Number of Eggs Laid in Clutch: | 5-8 |
| Days to Hatch: | 13-15 |
| Days Young in Nest: | 12-18 |
| Number of Broods Each Year: | 2, sometimes 3 |
| Lifespan: | 6 years |

*The house wren ranges from the Atlantic to the Pacific coast throughout Canada, the United States, and Mexico. It usually nests in the cavity of a tree or stump, in an old woodpecker hole, or in a bird box built for it. Its nesting materials include sticks and twigs, and an inner lining of hair, feathers, wool, spiders' cocoons, and catkins.*

### The "Look-in-on-Them" Nest Box

Edward H. Forbush, a New England ornithologist, made his first observational birdhouse some time in the 1870's, about the time he first began to attract birds. Forbush, probably the originator in America of many of our methods of bird attracting, built this simple nesting box to try to keep near his home a pair of chickadees that had been coming to his windowsill feeder all one winter. Forbush rabbeted the edges of the side of the box to receive a pane of glass *(B)* through which he hoped to observe the progress of egg-laying, hatching, and the growth of the young. Over the pane of glass he fitted a side panel, hinged at the bottom *(A)*, and held tight at the top by a hook and eyelet. This was only to be opened *(B)* for brief observations. Forbush mounted the chickadee nest box on a short board that he nailed to his windowsill and sat back to watch. The pair of chickadees came, accepted the nest box, and raised their families there for three successive years. In this way, Forbush was able to make scientific observations on the home life of chickadees, which might have been impractical or even impossible to make had they been nesting in a hollow in a tree.

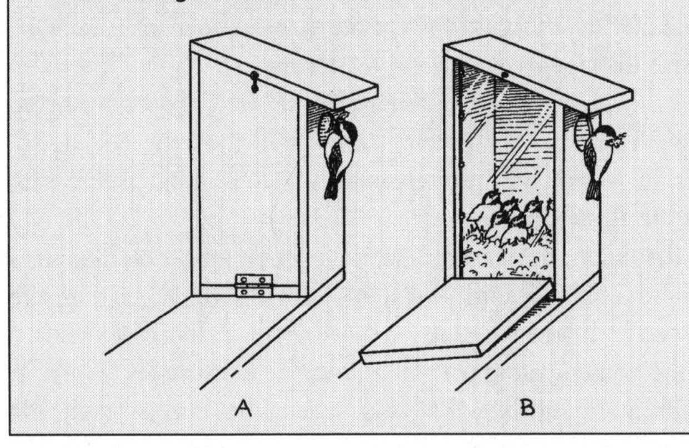


but both are interesting, attractive, and useful. Both use the same size nest box and will compete with each other for nesting places where they both occur. If you live within the range of the Bewick's wren or one of its many subspecies, you will surely want to put up nesting boxes to attract it to your yard.

The Carolina wren, which also has a distinct white stripe over each eye, is larger than the Bewick's wren, which it resembles. Like the Bewick's wren it sings beautifully and will often nest in bird boxes put up for it in our yards and gardens. Many years ago in a country place near Philadelphia, a pair of Carolina wrens entered the sitting room of a house through a window that was left partly open, and built their nest in the back of an upholstered sofa. They got inside the upholstery through a hole that had been torn in the covering material. The owner of the house was so delighted with his bird guests that he did not disturb them and they brought up their young successfully.

The Carolina wren is a more southern bird than the house wren and the Bewick's wren, although all three birds together live over certain areas of our country. In the eastern United States the Carolina wren seldom nests commonly north of New York City, although it has nested as far north as Maine. In the fall of 1949, I watched one of these birds flit about in a wooded thicket on Point Pelee, the southernmost tip of Ontario, Canada, where it had nested.

## BLUEBIRDS

In one of his many books, Enos Mills, the Rocky Mountains naturalist of several generations ago, told the story of his companionship with a family of bluebirds. These birds lived in and around his cabin for several years. Mr. Mills learned many things about these mountain bluebirds. One day he was delighted to discover that the five youngsters of the first brood, after they were able to fly, remained with the parent bluebirds to help feed their brothers and sisters from a second brood.

In the eastern United States, it is usually necessary to live outside of the limits of large cities to attract bluebirds. If you live in suburbs bordering open country you may get bluebirds, tree swallows, and purple martins to nest in your bird boxes. People who live in rural villages or on farms have the best chances of attracting them.

Of all our native birds, none is so widely loved as the bluebird—our true "bird of spring." With its sky blue back, reddish breast, and soft, sweet, warbling song, it has always been a favorite of country people. I remember from my childhood the thrill that my father and I got from discovering, not the first robin, but the first bluebird of the year. We always looked forward to its return to us on a mild day in March from the southern states where it winters.

**Bluebird Nest Boxes**

Make the round entrance hole of your bluebird nest boxes exactly 1½ inches in diameter. The slightly larger starling, the bluebird's strongest competitor for nesting places, cannot get into a bird box with an entrance hole this size. Build the box 5 inches square, inside dimensions, and 8 inches deep from top of the floor to the inside bottom of the roof. Bore the 1½-inch entrance hole 6 inches above the floor of the box. Chapter 3 explains how to build a "bluebird trail" using a variation of this basic box.

If you treat bluebirds with kindness and patience, these beautiful creatures may become very tame and friendly. A man I know of, who lived on the edge of a country town, trained a family of bluebirds, young and old, to be friends with his wife and children. He began by coaxing the birds to his windowsill for mealworms, of which they were very fond. Soon they learned to feed from his hands and to perch on his shoulders. Should you wish to try this yourself, mealworms, the immature or larval form of a beetle, can be bought in pet shops that sell cage birds and other animals.

During the past fifty years, old orchards, where bluebirds once nested so frequently, have almost disappeared. Either the old hollow apple trees have been pruned of their dead branches and their cavities filled, or the trees have been cut down. Farmers have replaced the old orchards with new young trees that they prune and spray regularly. This has improved the apple crop, but has discouraged crops of bluebirds, tree swallows, and other hole-nesting birds the old orchards once produced. Fortunately for these birds, as the old orchards disappeared, farmers and others started to put up nesting boxes for them. Bluebirds quickly adapted themselves to nesting in bird boxes, which may have saved them from disaster.

Bluebirds usually start to lay their eggs in April in the northern states, but in the South, they may have laid their first clutches of eggs by March. Three species of bluebirds nest across the North American continent—the eastern bluebird, mountain bluebird, and western bluebird. All will nest in bluebird boxes put up for them.

## TREE SWALLOWS

Along the Atlantic coast, this is the earliest swallow to move northward in spring. Many tree swallows winter in the South and arrive in the northeastern states in March and April. Like other hole-nesting birds, it has learned to rear its families in nesting boxes we put up for it. The tree swallow seems to prefer man-made nest boxes, although it hasn't altogether given up nesting in holes in trees and other natural cavities. If you have never put up nest boxes for them, you may not realize how badly they suffer from a "housing" shortage. At Cape Cod, Massachusetts, a scientist interested in birds wanted to see if he could attract more tree swallows in his area. Early one spring, he had ninety-eight nest boxes put up on posts and all ready for the swallows when they arrived from the South. That spring, his local nesting population of tree swallows rose from four pairs to sixty pairs, or fifteen times what it had been! To test the choice of tree swallows for various kinds of nesting environments, he put up boxes on posts in open fields, in the salt marshes, and in densely shaded woods. Of these different kinds of surroundings, the tree swallows much preferred to nest in the boxes put up for them in or along the edges of open fields.

Usually, pairs of tree swallows nest by themselves, somewhat away from other pairs of their kind. Although pairs occasionally nest close together in hollow trees, especially around marshes, it is best on small properties to put up only one or at most two nest boxes for these birds.

The entrance hole to the tree swallow box will admit bluebirds. The boxes built for these two kinds of birds are almost identical, so it isn't surprising that each will nest in the box of the other. This causes some competition between them, but both are very desirable birds to have in your yard and garden. When we lived in the

### Unwanted Visitors

To keep house (or "English") sparrows out of bluebird houses, place the bluebird house on a post *no higher than five feet above the ground*. House sparrows usually will not nest this close to the ground. If the house sparrows persist in nesting in the low-placed bluebird houses, cut a *rectangular* entrance hole in the box, 1½ inches high and only *1¼ inches* wide. This will allow the slenderer bluebird to get in, but will usually keep out the pudgier sparrow.

## Tree Swallow Nest Boxes

Make the entrance hole to the tree swallow box 1½ inches in diameter, the inside dimensions of the box 5 inches square, the depth, from top of floor to inside bottom of roof, 6 inches, and cut the circular entrance hole in the front panel any distance from 1 inch to 5 inches above the floor.

A more elaborate design is the Kinney tree swallow nest box designed by Henry E. Kinney of Massachusetts. Mr. Kinney's nest box is especially designed to protect young tree swallows from starving and from exposure during spells of rainy and cold spring weather when they need considerable insect food, which may be difficult for the parents to find. The front of the Kinney nest box has four holes in it; the largest is 1½ inches in diameter, which is the main entrance for the adults. The other three holes (each of *l-inch* diameter) and the entrance hole for the adults allow the young swallows to accept food without the parents' having to enter the nest box. This gives the adults more time to spend hunting insects. Mr. Kinney discovered that more nestlings survived in his nest box, and that more adults returned to nest in them in spring than in the standard tree swallow boxes with a single hole. The "T" perch nailed to the nest box gives the male tree swallow a perch on which he likes to stand guard close to the female, especially while she is incubating the eggs. The cleats on the roof are for the adults to cling to on windy days.

The scale of the detail drawing is ⅛ inch = 1 inch. Use wood ¾ inch thick. Drawing adapted from *Bulletin of the Massachusetts Audubon Society,* March 1952.

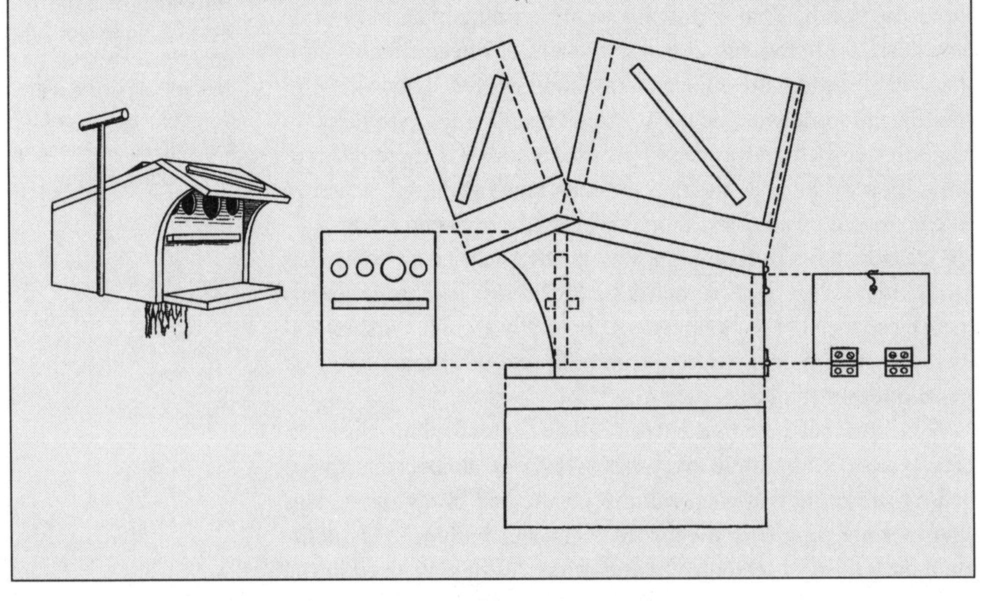

country, where we could attract these birds, we always allowed them to fight it out for possession of a nest box. We never interfered but we *did* try to be peacemakers by putting up extra nest boxes for them. Sometimes, if the tree swallows were driven out, they would turn to one of our extra boxes. At other times the bluebirds lost the fight and accepted one of the extra nesting boxes. For both tree swallows and bluebirds, we put their boxes on posts about 5 feet above the ground. To reduce the number of fights between these birds, we spaced their nest boxes at least 100 feet apart.

**TREE SWALLOW**

| | |
|---|---|
| Number of Eggs Laid in Clutch: | 4-6 |
| Days to Hatch: | 13-16 |
| Days Young in Nest: | 16-24 |
| Number of Broods Each Year: | 2 |
| Lifespan: | 5 to 9 years |

*The tree swallow ranges throughout the United States and Canada, wherever open water meets with marsh or meadow. Its nest, made of grasses and straws and lined with chicken feathers, is built in natural tree cavities, old woodpecker holes, birdhouses, rural mailboxes, and cavities in eaves or cornices of old buildings.*

## Violet-Green Swallows

This beautiful hole-nesting swallow lives over a wide area of the western United States, west of the Great Plains from Alaska to Mexico. It is as common in the mountains of Colorado as the house sparrow is in our eastern cities. Although it lives about houses and gardens of the West, it ranges far up the mountain slopes and into wild forest lands to build its nest. In Seattle a man built a nesting box for violet-green swallows and fastened it under the eaves of his house. For sixteen consecutive years after that, a pair of these birds nested in the box.

Make your nesting boxes for violet-green swallows of the same size and specifications as those for tree swallows.

## Barn Swallows and Cliff (or Eave) Swallows

When North America was a vast wilderness, barn swallows built their nests on the shelves of rocks in rock crevices or in caves. These two kinds of swallows, like the tree swallows, have learned to live in our yards and about our buildings, especially in country towns and on farms. Cliff swallows attached their bottle-shaped hollow mud nests to the walls of deep gorges in remote mountains and, sometimes, against the trunks of giant trees. In the summer of 1978, I saw a colony of cliff swallows at Lake Watson,

**CLIFF SWALLOW**

| | |
|---|---|
| Number of Eggs Laid in Clutch: | 3-6 |
| Days to Hatch: | 15-16 |
| Days Young in Nest: | 23 |
| Number of Broods Each Year: | 1 or 2 |
| Lifespan: | 4 to 5 years |

*The cliff swallow resides locally in Alaska and Canada, south to Mexico, from the Pacific to the Atlantic Ocean. It nests in colonies under eaves of houses, barns, churches, in canyon walls in mountains, or on a shelf built for it. The cliff swallow makes a flask shaped nest of mud or clay pellets, plastered against a vertical surface, with a side entrance.*

Colorado, that had built about 100 mud nests against the face of a red rock cliff, seventy-five feet up.

Today, the cliff swallow is often called "eave" swallow because it has almost deserted the cliffs where it usually built its nest. It now nests under the eaves of our barns and houses. Likewise, the barn swallow has gotten its name from its habit of nesting in barns. Usually there are never more than six to eight nests of this swallow in one building, but there are records of up to fifty-five in a large barn at Ipswich, Massachusetts, nearly all of which were occupied or "active" nests. These were built in the old-fashioned New England barns of many years ago when it was the custom to leave the doors open and swallows were free to fly inside to build their nests. Modern dairy barns and neatly painted farm buildings have closed doors and no open windows so that barn swallows need our help if we want to keep them around our homes.

## THE GREAT CLIFF SWALLOW EXPERIMENT

One of the most interesting projects to increase a "backyard" bird in this country succeeded brilliantly because of a chance discovery. At Deerfield, Wisconsin, the owner of a large barn had only one pair of cliff swallows nesting under its eaves in 1904. Appreciating the nesting requirements of these birds, he encouraged them to return the next year by nailing several strips of wood horizontally across the sides of the building to which they could more easily attach their mud-built, flask-shaped nests. The following season, several pairs of cliff swallows nested on the sides of his barn. Each year, the owner nailed up extra wooden strips to provide nesting places for more cliff swallows. Gradually he built up his colony until by 1911, several hundred of these birds returned to nest on his barn each spring. This seemed to be the greatest number of them that he could attract, for thereafter they did not increase.

Then came a wet year. After the cliff swallows had flown south for the winter, steady rains drove hard against the sides of the barn all that fall. The dried mud nests of the cliff swallows slowly crumbled and finally dissolved until not a one was left. The following spring when the swallows returned, each pair had to build a new nest, but instead of their usual numbers, almost twice as many nested on the barn. Was this purely chance—a happy coincidence? The barn owner decided to experiment.

Late that summer, after the swallows had left, *he knocked down every one of their nests*. When they returned the following spring, again they nested in greater numbers than ever before. By this time, the owner of the barn had the answer.

Each spring, previous to the wet year, many of the old swallow nests, instead of housing cliff swallows, had been invaded by house sparrows. Living in Deerfield throughout the year, these aggressive little birds got established in many of the old cliff swallow nests before the rightful owners returned in spring. But with the old nests gone, the house sparrows no longer drove a competing wedge into the swallow colony. They had to look elsewhere for nesting places and the cliff swallows could now build their nests and raise their young ones in peace.

### Barn and Cliff Swallow Nest Box

Barn and cliff (or eave) swallows do not nest in bird boxes. They need a horizontal ledge or some other support on the side of a building on which to build their nests. When I lived in the country where I could attract these birds, I fastened several rough, unplaned 2 inch by 4 inch joists, with the 4-inch sides flat against the building walls, on the outside of a small barn and of a garage in back of our house. I nailed these about 6 to 8 inches below the eaves to give the swallows that might nest there some protection from rain. The first year that I put them up, two pairs of barn swallows came and built their nests of mud, straw, and feathers on one of the joists.

In June 1942, more than 4,000 cliff swallows lived in the 2,011 nests they had built that spring on the sides of the Wisconsin barn. In 38 years, the owner had increased his cliff swallow population 2,000 times! As far as I know, this is the greatest backyard bird-attracting project to help a single species of bird in this country. It is also a model of patience and enduring kindness by the man who made it possible.

## PURPLE MARTINS

Each summer we went to Cape May, New Jersey, for a few weeks. In this quiet, old-fashioned resort on the southern New Jersey coast, I saw more colonies of purple martins nesting than I have seen elsewhere at one place in my lifetime. One spring day I counted more than 200 martin houses and most of these seemed to have martins nesting in them.

Almost every yard had its neat, white-painted martin house set up 15 to 20 feet above the lawn on a white post. Usually, glossy, blue-black martins were flying gracefully about over the grass or settling on the rooftops of these big birdhouses. Many of them perched and chattered conversationally on the porches or galleries that surround each of the floors of their "bird apartments." The rich, cheerful warbling of this largest of our swallows is one of the familiar sounds you will hear any summer day along the shaded streets of Cape May. Some of these martin colonies are so old that the birds that first came to nest in them have long since passed away.

One of the oldest colonies of martins I know of is in Greencastle, Pennsylvania. There, in the public square, in boxes put up for them by the townspeople, martins have been returning to nest since 1840. In more than a century, it is said, they have failed to return to Greencastle only during one period of time. For fifteen years after the Civil War the martins did not come back. To this day no one has been able to explain this mystery.

Martins become deeply attached to their nesting houses and

have a remarkable sense of accuracy in returning to them, which makes the Greencastle story even more puzzling. One spring, a man in Houston, Texas, had taken down his martin house to paint it. Some household chore diverted him from getting the martin house and its pole reset in his garden before his birds returned. Early one morning he heard a great chattering in his yard. When he went outside, his martins were flying in circles around the *exact* place in the air where their martin house should have been.

To keep house sparrows and starlings from nesting in the houses ahead of martins, some people don't put them up until the day that the martins usually arrive. A man in Massachusetts had just started to raise his martin house on its pole on the day he expected his birds to return. At that moment, some of his martins came flying into his yard. So eager were they to get into the house that they alighted on it and rode up with it while the man pulled and hauled it into place!

**PURPLE MARTIN**

| | |
|---|---|
| Number of Eggs Laid in Clutch: | 3-8 |
| Days to Hatch: | 16-18 |
| Days Young in Nest: | 26-31 |
| Number of Broods Each Year: | Usually 1 |
| Lifespan: | 4 to 8 years |

*The purple martin ranges from the Atlantic Ocean to the Pacific Ocean, from southern Canada south to Mexico. It nests in colonies in tree and cliff cavities, and mostly in birdhouses. Its nests are made of grasses, leaves, twigs, and feathers.*

## THE MARTIN HOUSE

A single-room martin house, built according to the basic nesting box design, 6 inches square and 6 inches high, inside dimensions, with a 2 1/2-inch entrance hole cut in the front panel 1 inch above the floor, will comfortably house one pair of martins. It will be more satisfactory, however, to get more than one pair of them in *one large birdhouse*, because they will nest together in colonies. The big, roomy "apartment" house can have 20, 30, or even 200 rooms, but an 8-room house will make an excellent start.

It is only fair to warn you that building a martin house is a bigger and costlier job than building the small, single-room houses that you may have already built for other birds. If you prefer to buy your martin house, a good one of 8 to 24 rooms will cost upwards of 100 dollars, depending upon how many rooms there are in the house and the quality of the materials with which it is built. Some suppliers are listed in the Appendix.

### Gourds Make Economical Martin Houses

In the spring of 1939, while driving along a country road in Georgia, I saw a colony of martins nesting in gourds. I had heard of gourds being used for birdhouses in the southern states, but this was the first time I had ever seen them in use. In the side yard, near a low plantation house, stood a white-painted pole 25 or 30 feet tall. From four X-shaped crosspieces nailed to the pole at different heights above the ground hung about 20 hollowed-out gourds, which are the fruits of plants closely related to the pumpkins and squashes. Each of the gourds had a pair of martins nesting in it. The variety most used for martin houses is the "bottle" gourd. Other varieties of the Lagenarias, or hard-shelled gourds, can also be used as houses for larger or smaller birds.

If you are interested in growing gourds to make your own birdhouses, the old government Farmers Bulletin 1849, *Useful and Ornamental Gourds* tells how and where gourds may be grown, and how to prepare them for birdhouses. Though no longer published, it is available at major university libraries that are repositories for U.S. government documents.

RULE OF THUMB

It's a good idea to verify that martins nest in your area before investing time and money in a martin house.

Before investing your time and money in a martin house, it would be desirable for you to inquire in your neighborhood and section of the state to see if martins have ever nested there, or if it is *probable* that they might nest in your community. In the Far West, I know of only two colonies of martins that learned to nest in birdhouses instead of in hollow trees. These were both in California, one in Loyalton, the other in Pasadena. Perhaps other colonies of martins have learned to nest in bird boxes on the Pacific coast, but they haven't done so like the martins of the eastern United States. Perhaps they will in time, particularly when more people put up martin houses.

If martins are already nesting close by, your chances of attracting them will be good, especially if you live near open fields and have a pond or stream. Even though your yard and garden may be ideally situated to lure martins, you might put up a house and still fail to attract them. If you are willing to risk the disappointment of not getting a colony of these birds, your neatly painted martin house, on its 15- to 20-foot-high pole in the center of your lawn or garden, will be a handsome accessory.

RULE OF THUMB

Painting your birdhouse white will ensure that your martins are comfortable during warm weather.

Start your martin house modestly, with one section of 8 rooms. Follow the directions for building it and for setting the pole on which to mount it as given in the illustrations on page 83. Build your martin house of the same kinds of wood and fittings recommended under "Building Materials for Birdhouses" (see box, page 60). Paint all martin houses white, including the pole to support it, and paint the roof and possibly the moldings around each section green or black. Either of these is an excellent color combination and makes a neat, attractive, and *cool* house because white paint reflects the sun's rays.

Some boys in a mining village in West Virginia had different ideas about color combinations for birdhouses. One boy painted his one-room martin house a bright yellow, another painted his red, another chose orange, and one boy painted his house red, white, and blue! Yet each boy had a pair of martins nesting in his house, which suggests that these birds may not be finicky about the colors of their homes. Nevertheless, be sure to paint your house white, if only for the comfort of the martins in warm weather. Martins like their houses well up in the air and in an open place where they get sunshine most of the day. In such a place, a large martin house, even with its central ventilating shaft and air space beneath the roof, can become very hot in summer.

## Martin House

Sections, or floor levels, may be built and added separately to the martin house *(A)* as the bird colony grows. A central air shaft and elevated roof allow air to circulate within and to cool the house. A molding strip around the underside of the roof and under each floor section *(see detail E)* holds them in alignment. The sections are held together by hooks and screw eyes *(E)*. Use 3/4-inch pine or other soft wood for walls and floors; use 1/2-inch wood for the roof and the interior partitions.

*Roof Detail B*—Roof with one side removed. The 6 inch by 6 inch opening of the central air shaft that should be cut out of the center of the attic floor is not shown. A 1-inch-high shoulder on the roof supports allows air to pass up under the eaves and into the attic, which has a ventilation hole at each end that is screened to prevent birds from getting in. The roof is 29 1/2 inches along the ridge top and 16 inches from peak to eaves level. The distance from outside to outside of the end ledges is 22 1/2 inches.

*Nesting Compartments, Detail C*—Make outside dimensions of floor 26 1/2 inches square. If you use one piece of wood for the floor, the metal angle irons shown will not be needed. Make each nest compartment 6 inches by 6 inches by 6 inches, inside measurements. Cut out the floor of the central chamber, which is the central air shaft. Make outside dimensions of nesting compartment 20 1/2 inches square, 3/4 inch thick. Use 1/2-inch pine for walls of inside of each nesting compartment. Cut the 2 1/2-inch entrance holes along a line 2 1/2 inches above the floor.

*Foundation Frame Support, Detail D*—This is the 20 1/2-inch square foundation, 2 1/2 inches high, for the bottom section. It has a center cross built of a double thickness of 3/4-inch oak boards, 2 1/2 inches wide. Attach four heavy angle irons, bolted to the oak frame as shown, which can be spaced to bolt to a supporting post 4 or 6 inches square, or to a round pole.

*Detail E of Porch*—When porch extension, instead of being part of floor, is attached separately, fasten with angle irons. Attach molding trim and angle irons with screws, as shown.

*Detail F of Pole Supports*—The 8-foot-long supports 4 inches square are set 4 feet deep in the ground. A heavy bolt or section of pipe serves as a hinge (G) and the base of the pole is held locked by two hardwood blocks or iron plates bolted together.

*Detail H of Pole*—Cross-section of a post 4 inches square, built of 7/8-inch hardwood. Supporting poles or posts for large martin houses should be 6 inches in diameter or 6 inches square.

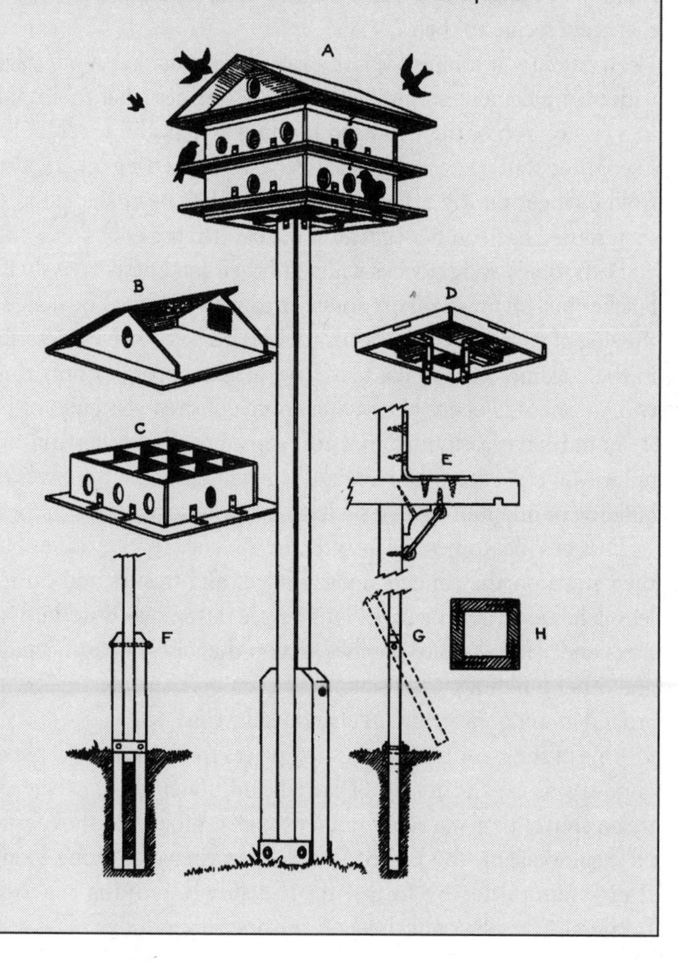

## ROBINS AND PHOEBES

Robins usually build their nests in trees, but like many other birds they are quick to use places that our civilization offers them. I have found their nests on fence posts, window ledges, the eaves of porches, and garages, on fire escapes, and other supports that seem secure to them.

Apparently a robin's idea of a safe place wouldn't always agree with your judgment and mine. I know of one nest that robins built in a city on top of the overhead running wire of the streetcar line. Every time that a streetcar passed beneath the setting bird, and the trolley wheel ran by a few inches below her nest, she stood up, then settled back on her eggs after the car had passed!

RULE OF THUMB

Be sure your phoebe and robin shelf has a sloping roof to keep rain off of nesting birds.

Both robins and phoebes will build their mud nests on a shelf or bracket put up for them on an arbor in your backyard or nailed to the side of your house, garage, or toolshed. These shelves, described in the diagram on page 85, may have all sides open, or only three, two, or one side open; but whatever type of shelf you build or buy, it should have a sloping roof to keep rain off the nesting bird, unless you can fasten the nesting shelf well up under the eaves of a building or other protective shelter.

Phoebes like to nest near water. In the country they often build their nests on the timbers under bridges and trestles and on rock ledges in caves and ravines. This gentle, attractive little bird also nests under the eaves of porches and on the joists of barns, garages, and other buildings. Occasionally it even builds its nest below the ground in mine shafts and in abandoned wells.

One of the most perilous nesting places that I ever saw a phoebe choose was on the inside of an I-beam under the carriage of a steam shovel that was idle for a few weeks. When the shovel started digging again, the parent phoebes continued to bring food to their young ones by following the slowly moving machine. Eventually they brought them off the nest successfully.

## Nesting Shelves

This simply designed nesting shelf for robins, phoebes, etc. should be built according to the size specified for each bird in the table.

Dimensions for Nesting Shelves with One or More Sides Open (See accompanying illustration.)

| Kind of Bird | Size of Floor (inches) | Depth of Bird Box (inches) | Height to Fasten Above Ground (feet) |
|---|---|---|---|
| Robin | 6 X 8 | 8 | 6-15 |
| Barn swallow | 6 X 6 | 6 | 8-12 |
| Song sparrow | 6 X 6 | 6 | 1-3 |
| Phoebe | 6 X 6 | 6 | 8-12 |

*Robins*—For the robin, make the nesting platform about 6 inches deep, from front to back, 8 inches wide, and 8 inches high from floor to inside bottom of roof. Drill several small holes in the bottom of it to drain off rainwater, just as you did with the bird nesting boxes. Place it anywhere from 6 feet to 15 feet above the ground.

*Phoebes*—Pattern the phoebe's nesting shelf after the robin's, but make it 6 inches square, inside dimensions, and 6 inches high from floor to inside bottom of roof. Place the phoebe shelf between 8 feet and 12 feet above the ground on the side of a building, arbor, or location similar to that of the robin's nesting platform. If you build your nest shelves, use the same kind of wood and other materials that you use for birdhouses, but it is better not to paint them. Leave the wood a natural finish.

## SCREECH OWLS AND SPARROW HAWKS (AMERICAN KESTRELS)

**SCREECH OWL**

| | |
|---|---|
| Number of Eggs Laid in Clutch: | 3-7 |
| Days to Hatch: | 21-30 |
| Days Young in Nest: | 25-30 |
| Number of Broods Each Year: | 1 |
| Lifespan: | 6 to 13 years |

*The screech owl, either red or gray, ranges from extreme southern Canada throughout the United States. It lays its eggs on the bottom of a natural cavity in trees, usually without nesting materials. It also nests in the old nesting holes of flickers, in sycamores, elms, utility poles, dead pines, oaks, desert saguaros, stream-side trees, and in birdhouses.*

Both of these interesting and useful birds will nest in boxes built for them, especially in suburban and even in city yards and gardens. Although these predatory birds usually feed on insects and wild mice, they occasionally kill and eat songbirds. This may make them objectionable in the small suburban garden, but there is no reason why screech owls and American kestrels shouldn't be invited to the country place. On a property of several acres or more, they should be encouraged to nest in bird boxes. I know of few birds that are more interesting and useful, and they should be welcomed as valued members of your bird community.

The American kestrel hunts largely for grasshoppers, crickets, and mice by day; the screech owl feeds upon beetles, moths, mice, and other creatures at night. Each feeds on almost the same *types* of food, except that one is on the day shift, the other the night shift. Both of them can enter a nesting hole down to about 3 inches in diameter, which means they are occasional competitors for nest sites.

One of the most interesting stories that I ever heard about these birds concerned a nesting hole in the dead limb of a eucalyptus tree in California. A pair of American kestrels were making frequent trips into the nest hole, obviously to feed their young ones. A man climbed the tree and found four young American kestrels and one young screech owl! The adult American kestrels were feeding the young screech owl, along with their own youngsters, and all of the young birds were in excellent condition. Possibly a screech owl had laid one egg in the nest hole and then had been dispossessed by the American kestrels, or the screech owl may have gotten to the nest of the kestrels and laid an egg in their absence.

At any rate, the kestrels had accepted their strange foster child and continued to feed and care for it until it too flew from the nest.

### Screech Owl and American Kestrel Nest Box

Build your screech owl and American kestrel nesting boxes exactly the same size. See the dimension table on page 55. Make the inside dimensions 8 inches square, and about 15 inches from the floor of the box to the inside bottom of the roof. Cut the 3-inch entrance hole in the front panel about 12 inches above the inside floor of the box. Drill small ventilation holes in the front panel *above* the entrance hole and make drainage holes in the bottom as prescribed for other nest boxes. It is better not to paint these boxes, so use a durable wood like cyprus. Fasten the nest boxes of these birds from 12 to 15 feet above the ground, preferably on the main trunk or on one of the upper vertical branches of a tree. An apple orchard is an excellent place to put up screech owl nest boxes. For American kestrels, locate the nest boxes in trees along the edge of a woodland or in a tree in an open field.

## DO BIRDS USE NEST BOXES AT NIGHT?

I don't like to disturb the birds that occupy our backyard nesting boxes, but we keep a close watch on their progress from the time that they lay their first egg until their last young one leaves the nest. We are eager to see them raise their families successfully, and so I look in the nest box occasionally to be sure that all is well with the eggs or young birds. I have done this without frightening the parent birds by waiting until I am sure that neither of them is in the nest box. Our boxes have hinged tops that lift up. It is easy to climb a stepladder and raise the roof of the box for a quick look inside while the adult birds are somewhere about, feeding themselves or gathering food for their youngsters.

One night I wanted to see which bird of a pair of flickers nesting in our yard spent the night in the nest box. At the time they had newly hatched young and either one or the other of the parent birds would be brooding them. Several hours after dark I took a flashlight and ladder, climbed to the box, and looked in. In the beam of the flashlight the sleeping flicker started, turned its head, and looked up. Quite plainly I could see the black mustachial marking on the side of its face. It was the male bird.

To see if the birds changed off in night brooding, I looked in the box on the following night. Again the male bird looked up at me. On successive nights, while the youngsters were small and required brooding, I always found the male flicker in the box. Later, when the young birds were well feathered and no longer needed the warmth of brooding, the male did not sleep in the nest box.

Long after the nesting season was over and the young flickers had gone away, I discovered what I believed was the same male flicker sleeping in the nest box at night. Flickers go to bed early, before most other birds. One evening, while I waited to see the male go to bed, a female flicker flew to the box and entered it. A few minutes later the male came. Cautiously he looked into the box before entering it, then in a fury he drove the female out. Before winter I put up an extra flicker box and saw a female enter it one evening about an hour before dark. A few weeks later when I looked into the box at night, I found that another male flicker had taken possession of it.

## *Roosting Shelters for Hole-Nesting Birds*

Hole-nesting birds need roosting shelters, particularly on cold winter nights. I once read about a man in the state of Washington who had put up a 6 inch by 6 inch birdhouse in his yard in which a pair of violet-green swallows nested in summer. One evening in fall, after the swallows had gone

## The Roosting Box

Make bird roosting boxes of the same kind of wood, hinges, screws, and other materials recommended for nesting boxes. For small birds, make the roost box about 10 inches square, inside dimensions, and about 3 feet high from floor to roof. Have a "lift-up" top *(A)* for periodic inspections or cleanings. Cut the 3-inch entrance hole in the front panel *no more than 2 inches above the floor of the box.* This will allow the accumulated warmth from the bodies of birds sleeping in the box to rise to the top and to be held there. If the entrance hole in the roost box were cut high up near the roof, as in nesting boxes, the body heat of the roosting birds would be lost and the birds less able to keep warm.

Make round perches about 1/4 inch in diameter for the small birds to roost on, and fasten them with carpenter's glue into small holes drilled into the inner walls of the roost box. Stagger, or offset, the perches as shown in the illustration to prevent birds from roosting directly above or below one another. The open section of the box *(B)* shows the placement of the roost perches inside. It is better not to paint your roosting boxes.

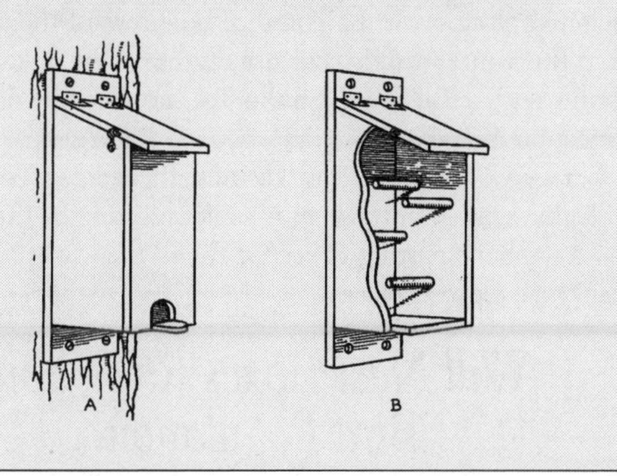


south for the winter, he saw thirty-one winter wrens crowd into this bird box to spend the night! A woman in Nebraska who attracted birds kept nesting boxes up for their use all year round.

After a heavy February snowfall, she watched one evening while eight bluebirds entered a flicker box, one by one. The ninth and tenth birds could not squeeze into the box and had to leave the garden to find a roost elsewhere on that cold and stormy night.

A safe warm place to sleep may save the lives of many birds that are wintering in your backyard. Although our birdhouses give roosting shelter to screech owls, flickers, downy woodpeckers, nuthatches, chickadees, wrens, and other birds, there is not always room in them to accommodate all the birds that try to use them. Even blue jays and certain others that do not nest in bird boxes will sleep inside a roosting shelter.

In the fall we put up four roost boxes in addition to our eight birdhouses that are always in place and available to birds both in summer and in winter. These roost boxes are easy to build and are designed to make use of the body heat of the birds themselves to keep the boxes warm.

Place your roosting box 8 to 10 feet above the ground on a tree or post, as recommended for birdhouses. Be sure that it is in a sheltered place, with the entrance hole toward the south, away from the winter winds that usually blow out of the north or northwest. Use cat guards on the trees or posts like those used to protect birdhouses.

Screech owls, American kestrels, flickers, and other larger birds that nest in holes seem to prefer roosting by themselves. If you have built nesting boxes for them, these will also provide them with roosting places.

## *Will Nest Boxes Raise Your Bird Population?*

In 1926, at Bell, Maryland, W. L. McAtee, a government naturalist, started a fascinating experiment to increase songbird populations. Our federal government had planted Asiatic chestnut trees that showed a remarkable resistance to the blight, an incurable fungus disease that had doomed most of our American chestnut trees. Even though the Asiatic trees were immune to blight, the *chestnuts* were not safe from the attacks of an insect called the nut weevil, which sometimes eliminated 50 percent of the chestnut crop. The government needed just as many sound chestnuts as they could reap to perpetuate this promising tree. What would destroy the nut weevils?

Birds would help, because at least eighty-four kinds feed on nut weevils. Among these, downy woodpeckers, chickadees, titmice, crested flycatchers, and other hole-nesting birds that eat nut weevils might be lured to the chestnut tree planting by putting up nest boxes for them. McAtee, an expert in attracting birds, put up a sixteen-room martin house, a large birdbath, and ninety-eight nesting boxes on the 3½-acre tract, or about twenty-eight nest boxes on each acre of land. The boxes were kept in place for about six years and a careful record kept of the kinds of birds that nested in each box and how many broods of young they raised. Eight different kinds of hole-nesting birds—starlings, house wrens, house sparrows, purple martins, bluebirds, crested flycatchers, flickers, and tufted titmice—came to live in the boxes. Although the birds did not completely eliminate the nut weevils, they helped to control them and added considerably to the bird life of the area. Just the adult hole-nesting birds, without counting the hundreds of broods of nestlings they raised, swelled the local bird population by about four times.

You may duplicate McAtee's accomplishment on your own property and you might achieve an even greater increase of birds. A woman in Minnesota put up nesting boxes and, by attracting a large colony of martins and other hole-nesting birds, raised her bird population to two hundred nesting pairs on 3½ acres of land. This is about 57 pairs an acre, which is almost a record for attracting songbirds in this country. There are many ways by which you can attract, in large numbers, an interesting variety of useful birds to your yard and garden.

**BLACK-HEADED GROSBEAK**

| | |
|---|---|
| Number of Eggs Laid in Clutch: | 2-5 |
| Days to Hatch: | 12-13 |
| Days Young in Nest: | 12 |
| Number of Broods Each Year: | 1 |
| Lifespan: | 5 to 6 years |

*Black-headed grosbeaks range from the Pacific Ocean to the Rockies. Their loose, thin nests are made of rootlets and twigs and are usually found 6 to 12 feet up in the dense outer foliage of deciduous trees or shrubs along streams and also in gardens.*

# *Building a Bluebird Trail*

had known that our eastern bluebird, symbol of happiness and eternal spring, had been declining in numbers for seventy-five years; records kept by American ornithologists and bird-watchers had proved that. But I was not prepared for the shocking news that came after the winter of 1957-58. During that season of unusually cold weather in the South—called by bird-watchers "the year of disaster"—bluebirds, robins, hermit thrushes, phoebes, and other songbirds that winter there had starved or were frozen to death by the thousands.

Bluebirds were hardest hit. Most birds can stand unbelievable cold if they are well fed. But with the ground and trees coated with ice or snow through much of the winter, the bluebirds' foods of wild berries and insects were unavailable to them. People found their frozen bodies scattered over the ground from Ohio east to Virginia, and from Tennessee through the Carolinas to Florida, the eastern bluebird's main wintering areas. An estimated one-third to one-half of the entire eastern bluebird population was destroyed. Perhaps most of the survivors were those that had learned to come to bird feeders for their favored dried currants, raisins, and chopped peanuts. At Chapel Hill, North Carolina, I had found I could help them in winter by giving them a doughy paste that they loved; I mixed bacon drippings or suet with peanut butter, cornmeal, and flour and stuck it in the rough bark of trees or put it in my suspended bird-feeding sticks.

I knew that feeding bluebirds in winter could help them, but it could not overcome the widespread diminishing effect on their numbers due to the six harsh winters that followed the disastrous winter of 1957-58. Unusual winter cold in the South continued to kill bluebirds until their population reached its lowest point in history during the spring of 1963. People who loved them and who watched them return in thinning numbers in the spring were aghast. How long could the bluebird, so vulnerable in winter, survive? I decided to help them in the most effective way they could

**EASTERN BLUEBIRD**

| | |
|---|---|
| Number of Eggs Laid in Clutch: | 3-6 |
| Days to Hatch: | 13-16 |
| Days Young in Nest: | 15-16 |
| Number of Broods Each Year: | 2, sometimes 3 |
| Lifespan: | 3 to 7 years |

*Eastern bluebirds range across eastern North America and west to the Rockies. Their grassy nests are built in the natural cavity of a tree or post, or in a woodpecker hole. Bluebirds will also nest in birdhouses.*

be helped; along with others, who were doing something about their plight, I would build a bluebird trail.

Bluebirds do not nest in the open in a bush or a tree as many birds do. They carry nesting materials of soft grasses and fragrant pine needles into the abandoned nest holes of woodpeckers, or into the dark hollows of trees and old wooden fences. But most of these favored places are gone. Farmers have replaced the old orchard trees with younger, neatly tended trees without hollows; most fences are no longer wooden but are made of steel posts and wire.

Suffering from an acute housing shortage, bluebirds have been further deprived of nesting places by the more aggressive European starling and house sparrow—introduced into North America in the latter half of the nineteenth century—which also

### Dangerous Parasites

The often reported deaths of nestling bluebirds, tree swallows, and house wrens, still in the bird nesting box and unable to fly, may be caused especially when the bodies of these young birds are heavily parasitized by the blood-sucking grubs (larvae) of the bluebottle fly (see suggested methods of control, pages 241-243). Raymond F. Potter, a bird-bander at Enfield, Maine, tested a new method for control of the prevalent larvae of the bluebottle fly in bird boxes. In thirty-four of the thirty-seven tree swallow nesting boxes, after the adults had laid their eggs, he dusted both eggs and nest with 1 percent rotenone. Meanwhile, for comparison, Potter had not treated three of the tree swallow nesting boxes. Nesting success for each spring (young raised to flying stage or leaving the nest) was 100 percent in the treated nest boxes but only 69 percent in the untreated ones. Potter concluded that using rotenone in bird nesting boxes in spring not only controlled the larvae of the bluebottle fly, but the larvae of other parasites in birds' nests such as bird fleas and tachinid (biting) flies. But be careful that rotenone does not get into a pond or stream, for it destroys aquatic life.

nest in cavities. If anyone was to help the bluebird, he would not only have to build bluebird houses for the garden but also to make many boxes available to bluebirds in the parts of the countryside where most of them live. And the boxes would have to be designed and hung in a way that would discourage preemption of them by starlings and sparrows.

I first heard of bluebird trails from my friend and adviser, T. E. Musselman of Quincy, Illinois. In 1934, he had set up 25 bluebird nesting boxes along country fence posts in Adams County, Illinois, a place in which bluebirds had almost disappeared. In February 1935 he added more houses until 102 were placed along forty-three miles of country roads. Eighty-eight of the boxes, or 86 percent, were quickly occupied by bluebirds. In the published report of this conservation project, Dr. Musselman wrote that "for the first time in twenty years, bluebirds are a common sight along the roads of Adams County, Illinois, and I believe that any other enthusiast can duplicate this." He called his series of bluebird nest boxes a "bluebird trail," and is credited with originating both the name and the idea.

One evening I went before the Orange County Boy Scout Council, a group of businessmen, doctors, lawyers, writers, and others in Chapel Hill who serve the local Boy Scouts and Cub Scouts by overseeing their projects. I needed at least thirty bluebird houses to start a bluebird trail, and I thought the youngsters could help. And what better project could one offer children? It would not only help bluebirds, but it would give the youngsters an opportunity when traveling the bluebird trail to learn about birds and their ways.

RULE OF THUMB

**Mount your bluebird boxes about 4 feet above the ground, facing a quiet road, but away from pastures where grazing animals might damage them.**

The council agreed and the Boy Scouts and Cub Scouts built sixty bluebird nesting boxes from a sample bluebird house I had given them. Under my instruction, they wired or nailed them to fence posts along farm-country roadsides. The boxes were set about 4 feet above the ground, each one facing the road, away from pastures where grazing cattle might rub against the boxes and break them. We placed the boxes about 400 feet apart. Setting them closer together would result in too much fighting for their possession (bluebirds are strongly territorial and will drive out other bluebirds from their chosen nesting areas).

**Don't Worry the Bluebirds**

After checking on eggs laid by bluebirds, leave the nest box quickly. Bluebirds lay one egg a day and will not remain in the box incubating until the full set of eggs is laid.

The memory of the first bluebird house we hung one April day on a fence post at the edge of a farm field will linger in the mind of every boy who was there. One of the boys, "for luck" he said, had scratched with his knife Home Sweet Home on the front panel or door of the box. We had nailed the box to the post; it was low enough so that a house sparrow would hesitate to nest in it, and it had a 1½-inch round entrance hole, which was too small for a starling to enter.

As we walked away and stood at the edge of the road to admire our work, a handsome male bluebird flew to the entrance hole and clung there, looking inside. He warbled gently to his mate. When she answered and flew to him from a tree, we saw that she was a paler blue and a softer brown than he. She alighted at the entrance hole, looked in the box, then shivered her wings ecstatically. We got in our car and drove on, fearful of frightening her and wondering if she would accept the new brown-painted home we had offered her.

On the way back from our three-mile trip down the road to hang the remaining bluebird houses, we stopped near the first nesting box. The boys shouted at what they saw. The female bluebird, with some brown pine needles in her bill. was entering the box. Five days later, she had completed the nest; two days later, she had laid her first pale blue egg. After that, we discovered that she laid an egg each day for four days, usually in the morning between seven and eight o'clock.

When our bluebird had laid her five eggs, she began to incubate them, spending long hours inside the box with her warm breast pressed close to the eggs. Rarely did she come outside, but the male came regularly to the entrance hole with a grasshopper, a beetle, or a soft caterpillar to feed her. Each time that he arrived, we saw her head appear at the doorway to take food from his bill. Then she disappeared to settle down again on her nest.

Two of the boys went with me once a week to inspect the bluebird trail. After fourteen days the eggs had hatched in the first nesting box, and, within another two weeks, the five young bluebirds had flown. By the middle of June, thirty of our bluebird houses had bluebirds nesting in them. Others of the houses had nesting chickadees, wrens, titmice, or nuthatches in them, especially those boxes we had placed on fence posts near woods. These smaller birds were welcome too and so were the big-eyed flying squirrels and the small deer mice that built their nests in two of the empty nesting boxes.

By summer's end, our pairs of bluebirds had each raised two or three broods, but not without loss. Several broods disappeared when they were very small, possibly eaten by a snake or a raccoon. A farm cat caught one young bluebird just after it had flown from the box. But from our regular visits along the trail, we knew that our sturdily built boxes had sent more than two hundred young bluebirds successfully into the world.

The next spring we added more bluebird boxes to the trail. By August—the end of the nesting season—under their parents' care, more than three hundred young bluebirds flew from our seventy-five nesting boxes. With the passing of the next two winters, bluebirds in our part of North Carolina were almost as common as they had been before the cold winters of the previous decade.

From time to time I had received similarly encouraging reports from a retired doctor, and several each from educators, insurance men, engineers, construction workers, and housewives who have built bluebird trails. Stiles Thomas of Allendale, New Jersey, began his first bluebird trail in 1955 when he and other members of the Fyke Nature Association put 69 bluebird nest boxes on country fence posts and poles. Within five years they had placed 107 boxes, of which the most successful were those attached to fence posts and utility poles, rather than to trees.

### Clean Your Nest Boxes

Many bluebirds will not nest in a box that has old nesting material in it. After each nesting is over and the young birds have left the box, sweep out the old nesting material. Be certain that the nest is abandoned and not a newly built one.

### Disinfecting and Repairing Birdhouses

When you clean out your birdhouses, just before the nesting season, this is an excellent time to repair or replace sides, tops, or bottoms that may be cracked, broken, or rotted from long use. If you think the outside of the bird box needs to be treated again with a wood preservative or repainted, this is a good time to do it. After cleaning the inside of the box thoroughly, spray with creolin (1 part creolin to 10 parts water), obtainable at most drugstores. This will destroy bird lice or other insects that may remain hidden in the cracks and corners.

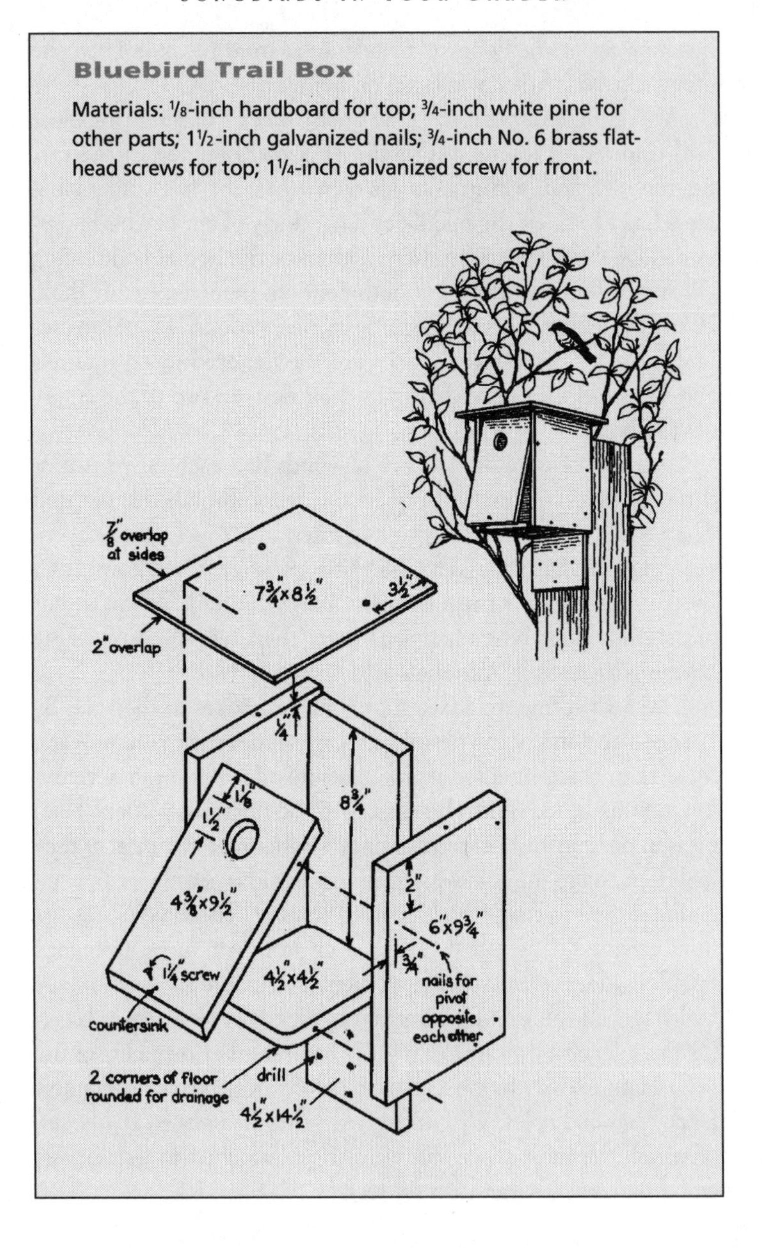

### Bluebird Trail Box

Materials: 1/8-inch hardboard for top; 3/4-inch white pine for other parts; 1 1/2-inch galvanized nails; 3/4-inch No. 6 brass flathead screws for top; 1 1/4-inch galvanized screw for front.

## How to Build a Bluebird Trail

Bear in mind that not only bluebirds but chickadees, nuthatches, titmice, wrens, and other desirable hole-nesting birds will also nest in your bluebird houses. These are good tenants just as useful in their insect-eating habits as bluebirds. When you tell farmers and other landowners about your purpose—that you are trying to help birds that in turn help them—I doubt if anyone will refuse you permission to put up boxes on his land.

From my own experience with bluebird trails, and from information I have gathered from those who have long had them, here are some good rules to follow:

1. Choose as a trail leader a parent, Scout troop leader, or other responsible person. Start a bluebird trail at any time between February 1 and July 1; however, it is best to have boxes up early for first nestings, which may be as early as March. Some bluebirds have second and third nestings through May, June, and into July. The nesting range of the eastern bluebird is from Newfoundland south to Florida and Bermuda, west to the Rocky Mountains, and from Saskatchewan south to Texas. The nesting range of the mountain bluebird is, as the name implies, in high mountains (mostly 5,000 feet) in summer from Alaska and Manitoba, south to the mountains of California, Nevada, Arizona, New Mexico, Oklahoma, Colorado, Nebraska, and the Dakotas. The western bluebird is resident in summer and winter from south British Columbia and central Montana, in mountains south to northern Baja California.

2. It is best to start modestly, with twelve to fifteen bluebird nest boxes. Placed four to a mile, these will stretch three or four miles out of town. You can add more, or start a new trail each year.

3. Paint the boxes (white, gray, brown, or green) before you put them up, and paint them on the outside only.

4. Choose a quiet secondary road where there are lots of fence posts and open fields, and occasional woods along the way. This is typical bluebird country—they like open fields adjacent to country roads and one can often see them perched on utility wires by the roadside. Some bluebird trailers prefer roads

**SUMMER TANAGER**

| | |
|---|---|
| Number of Eggs Laid in Clutch: | 3-5 |
| Days to Hatch: | Probably 12 |
| Number of Broods Each Year: | 1, possibly 2 |
| Lifespan: | 6 years |

*The summer tanager ranges from Delaware and Maryland south to Florida and west to Oklahoma, Nebraska, and southern California. Its nest is a loose shallow cup of leaves, weed stems, and grasses built 10 to 35 feet up, well out on the horizontal limb of a dogwood, oak, or pine in an oak-pine wood. In the Southwest, the summer tanager nests in a willow or cottonwood.*

**VIOLET-GREEN SWALLOW**

| | |
|---|---|
| Number of Eggs Laid in Clutch: | 4-6 |
| Days to Hatch: | 13-16 |
| Days Young in Nest: | 23 |
| Number of Broods Each Year: | Usually 1, possibly 2 |
| Lifespan: | 5 to 9 years |

*The violet-green swallow ranges west of the Great Plains, from Alaska to Mexico. Its nest is built of weed stems, grasses, and feathers, in colonies, in holes of dead pines, in deserted woodpecker holes, birdhouses, or in the natural cavities in cliffs.*

that make a circle and will bring them back to the town from which they started. Others do not mind backtracking. You may start your trail in your garden, or in the garden of a friend, and extend outward, unless of course you live in a city.

5. Ask permission of landowners to put up bluebird boxes on their fence posts. Nail or screw each bluebird house to the side of the fence post that faces the road. Besides nails or screws, carry some wire with which to fasten the bluebird houses to metal fence posts.

6. Set the boxes no more than 4 feet above the ground. House sparrows are less likely to occupy a low-placed nest box; bluebirds will accept them. I have seen bluebirds nesting in the hollows of wooden fence posts no more than 3 feet above the ground, but about 4 to 5 feet is a good height—about one's eye level with the entrance hole of the box—as this allows easy inspection of the nesting progress when opening the swing-front panel of the box. *Do not* put up boxes on fence posts near farmhouses, as house sparrows usually congregate there and will occupy them to the exclusion of bluebirds.

7. Some authorities put up no more than four nest boxes per mile but I have found, as have others, that pairs of bluebirds do not always have territorial squabbles if the boxes are closer together. In your yard you may place up to three bluebird houses, but put them on different sides of the house. This prevents each nesting pair from seeing the others at their nest boxes and reduces fighting. The boxes on our bluebird trail at Chapel Hill were put on fence posts about 400 to 500 feet apart. We had about 80 percent occupancy by bluebirds our first year and had comparable nesting occupancy in the years that followed. This is equal to or better than average.

8. Boxes may also be attached to isolated trees, trees at the border of open woods, and trees of small or medium size, as bluebirds do not seem to like boxes on large trees—perhaps they fear the squirrels that often race up and down a large tree in their hunt for nuts and acorns. Boxes may also be attached to utility (electric power and telephone) poles, but

unless company permission is secured, these may be torn down by maintenance linemen.

9. Inspect your bluebird trail at least once a week. Early on Sunday morning is often a good time. Much of the fun and rewards of the trail are in these periodic inspections. The trail leader should open the nest boxes—tap on the box first to allow the bluebird on the nest to get out before you open the box. My friends, especially the young ones, are amused when I tap on the box, or knock on the side panel and ask, "Is there anybody home?" The swing-front panel of the Dick Irwin Bluebird House is ideal for quick and easy inspection. Open the set-screw at the bottom of the front panel with a screwdriver. After examining the contents for nest-building progress, number of eggs, or growth of the young, make notes for later reference, then swing the panel back in place and screw it shut to prevent an easy opening of the box by people who might not be as interested in the welfare of bluebirds as you are. By taking notes on eggs and young, you will learn how long it takes for the nest to be built, eggs to hatch, and young to grow large enough to leave the nest.

10. Maintain your bluebird trails—repair or replace boxes damaged by accident or vandals. A small white card tacked to the front of the box below the entrance hole, telling what the box is, and its purpose, may help prevent the destruction of the box by thoughtless people. We tacked the following message on the front of each bluebird box of the bluebird trail at Chapel Hill:

THE DICK IRWIN BLUEBIRD TRAIL
Please do not disturb
This box was designed by Dick Irwin of Anchorage, Kentucky, to help give bluebirds a place to nest. One spring day, while many bluebirds were building their nests in his boxes along roads in Kentucky, Dick Irwin died at age 64. This trail is dedicated to his memory.

Not a single one of our boxes was destroyed.

W. G. (Bill) Duncan of Louisville, Kentucky, not only developed his own bluebird trail, but also supplied people throughout the United States and Canada with bluebird houses of his own successful design. He also originated a "Bluebird Letter," whose stimulating messages went out to hundreds of correspondents.

William Highhouse of Warren, Pennsylvania, encouraged by Bill Duncan, began a bluebird trail in his county in 1957. Within five years he had placed 100 bluebird nest boxes, about 3 to a mile, along rural roads. From these, eastern bluebirds raised more than a thousand young, of which almost eight hundred were from the first (usually more successful) nesting season. His bluebird trail became a full-time hobby.

Inspired by the work of Bill Duncan, Dick Irwin of Anchorage, Kentucky, put up 161 bluebird boxes on fence posts over seven square miles of Kentucky roadsides beginning in the spring of 1961. In 1962, his sturdily built nest boxes had seventy-two pairs of bluebirds nesting in them; they produced two hundred young.

His nest boxes are the most practical I have ever seen. The boxes are 4 1/2 by 4 1/2 by 9 1/2 inches, with a 1 1/2-inch entrance hole in a front panel, which, for regular inspection of nesting, swings open on two nails driven into the sides of the box. (No hinges are needed.) They are tightly built, warm and dry inside, with a ventilating space just under the roof (see illustration, page 98).

Dick said that his bluebird boxes, if kept painted (he painted them white, which reflects light and keeps them cooler) will last fifteen to twenty years. He sent me 6 to get me started and these began the Dick Irwin Bluebird Trail of Chapel Hill, North Carolina—the trail that I had started with the Boy Scouts and Cub Scouts and that I dedicated to him.

# *Helping the Birds at Nesting Time*

## A Bird's Need for Nesting Material

Before I started to attract birds, I never thought they might suffer from a shortage of nesting materials. Perhaps most birds don't ordinarily, but the story of the materials that a house wren once put in her nest would make you believe so.
After the nesting season was over, a curious woman living in Ames, Iowa, examined a house wren's nest in her yard. She found the following items, which she separated and counted:
Fifty-two hairpins, 68 large nails, 120 small nails, 4 tacks, staples, 10 straight pins, 4 pieces of pencil lead, 11 safety pins, 6 paper clips, 52 pieces of wire, 1 shoe buckle, 2 fishhooks, and 3 garter fasteners!
This pair of house wrens seemed to have had a taste for hardware, but their choice was unusual. We can supply birds with natural nesting materials and we can have a lot of fun doing it.

The more I watch birds, the more some of them astonish me with their adaptability. In human beings we might describe this as a practical quality through which certain people always "make the best of things." In the same way our house wrens, even if our backyard were completely bare of plants, would probably pick up nails, pins, wires, or any other odds and ends with which to build their nests, provided we put up nesting boxes for them. Fortunately we have trees, shrubs, and a lawn, and from these the wrens gather the coarse dead twigs and dried grasses they ordinarily use to build their nests.

When our house wrens started to build, we would often pick dead twigs from our spirea and dogwood bushes and put them in neat little piles on the lawn. The sharp-eyed wrens soon found them and carried them hastily into the nest box, as if they feared that some other bird might find this treasure and make off with it before they did.

Once, to see if they always preferred dry, dead twigs, we mixed in a few fresh green ones from which we had stripped the leaves. One of the wrens immediately picked these out of the pile, one at a time, flew to a corner of the yard, and dropped them. Then both birds went to work on the remaining dead twigs and quickly carried them into the nest box. Apparently they rejected the green twigs because a nest built of these would undoubtedly make the inside of the nest box damp. Such is the guiding instinct of birds.

We once had a pair of northern orioles that nested either in our suburban New York City yard or in our neighbor's yard every year. They didn't find on our properties the milkweed plants from which they usually gather long fibers to weave into their nests. Instead of going on a long hunt for these materials they learned to use bits of string, cotton yarn, and strips of cloth that we put out for them. The instincts of these adaptable birds have told them that our gifts are acceptable substitutes.

## A Bird Tragedy

One spring, when I first started to experiment with the willingness of birds to accept man-made nest materials, I was living on a farm in upstate New York. A pair of northern orioles were building a nest in the drooping end-branches of an elm tree in my front yard. Each morning, I put out strings, finely torn strips of rags, and cotton waste on the lawn for them, and by noon, the orioles had taken every bit of it to their growing nest. By the time they had completed it, they had used about seventy-five feet of white string, which I had cut into twelve-inch lengths for them.

Then came the tragedy. One afternoon I looked up in the tree and saw the bright yellow and black body of the male bird hanging limply from the side of the nest. With an extension ladder I reached him and found that he had gotten his neck entangled in one of the strings. The string, like all those I had cut for the orioles, was so long that it had not been completely woven into the nest. A loop in it had trapped the bird. He was dead—strangled to death by one of my good intentions. I had heard of orioles, robins, chipping sparrows, swallows, and other birds dying from getting enmeshed in horsehairs, plant fibers, and other natural nesting materials, but I never quite understood how, until I saw the treacherous loops in those long strings. Since then I never put out string, yarn, flax, thread, horsehairs, or strips of cloth any longer than about six to eight inches. As far as I know, none of the birds that have used them have been injured.

WARNING

To avoid the risk of entanglement, never offer your birds any piece of nesting material longer than six to eight inches.

# How to Offer Nesting Material to Birds

RULE OF THUMB

The safest places to hang nesting materials are on the limbs of trees and shrubbery.

One hot July morning a few years ago we cut short pieces of white and dark-colored yarns and simply dropped them on the grass in the center of our lawn. A female robin, building a nest for her second brood that summer, quickly discovered the pieces and carried them away as fast as we put them out for her. The next day, we dropped short pieces of cotton twine on the lawn and watched her carry these to her nest.

Because it is safer for birds, we ordinarily hang nesting materials on the limbs of trees and shrubbery. There, robins, orioles, and vireos can pick these up without being stalked, and perhaps caught, by cats that might sneak up on them while they were absorbed in gathering nest materials from the ground.

Some of our friends prefer to put the stringlike materials in a small wire basket hung from the branch of a tree, or to nail a wooden container to a tree trunk and fill it with nesting material. Later, the wire basket and wooden container can be used to hold suet in winter.

We have taken the mesh bags that oranges are sold in, and have filled these with yarn, strips of cloth, horsehair from old upholstery, bristles from worn-out paintbrushes, cotton batting, pieces of wool, nonabsorbent cotton, chicken feathers, pieces of soft cloth, rags, paper, bits of fur, excelsior, and dried grasses. We have watched robins, orioles, yellow warblers, chipping sparrows, song sparrows, goldfinches, cedar waxwings, and other birds come to the bag to make their own choice of nest materials from this hodgepodge.

We try to keep these bags or other containers filled with nest materials available to birds from May through August. Some birds, like the robin, may build two, or possibly three, nests during the summer. Others—goldfinches and cedar waxwings—do not nest until late summer, usually in July and August. Certain other birds require even different nest materials, which we also offer them.

## MUD FOR BIRDS

**TRY THIS**

You can also offer birds nesting materials in small wire baskets, mesh bags, or wooden containers attached to trees.

One of the queerest robin nests that I ever heard of was built in a tree in a Boston backyard in the early 1900s. The female robin usually builds most of the nest, and the male helps by bringing her materials. This female, with an extraordinarily feminine touch, had woven into her nest two red satin ribbons worn by people who attended the 1903 convention of the National Education Association. Trailing partway out the side of the nest, the gold lettering on one of these ribbons read plainly:

New York N.E.A. at Boston, 1903

Near the nest rim the bird had embedded a piece of coarse white lace through which she had neatly threaded two white chicken feathers. The remainder of the nest she had decorated with brown string, yellow string, a piece of blue embroidered silk, the hem of a handkerchief, and a bit of white satin ribbon! But inside the nest she had unfailingly built a hidden lining of mud. Without mud, no robin nest is complete.

Mud is so important to some birds in nest-building, particularly to cliff and barn swallows, that it may be impossible for a colony to remain established without a dependable supply. The man who encouraged the great cliff swallow colony to nest on his barn in Wisconsin kept large pools of water and mud near the barn all during the nesting season. He claimed that the swallows could not have built the 2,000-nest colony without this large and reliable mud supply.

**SCRUB JAY**

| | |
|---|---|
| Number of Eggs Laid in Clutch: | 2-7 |
| Days to Hatch: | 16-19 |
| Days Young in Nest: | 18 |
| Number of Broods Each Year: | 1 |
| Lifespan: | 11 years |

*The scrub jay ranges from western California north to Washington, east to Colorado, and south to Texas, also in Florida. It makes its bulky nest out of sticks and rootlets in a low tree or bush, 2 to 12 feet above the ground.*

## FEATHERS FOR SWALLOWS

One spring morning, a woman in a New Hampshire farmhouse sat at an upstairs window looking out over the sunlit fields. She had been very ill all winter and her recovery had been discouragingly slow. As she looked gloomily down into the farmyard, a breeze picked up a white chicken feather and whirled it into the air. Higher and higher it hoisted the feather until it had lifted it to the height of the woman's window. There the feather floated motionless, but only for a second. Suddenly an arrow-shaped steel blue form struck it, darted toward the ground, rebounded high into the air, and with the feather in its beak raced toward the barn a hundred yards away.

The woman gasped with admiration. A barn swallow had snapped the feather out of the air and without a pause had flown swiftly with it to the barn and through the big open doors. Somewhere within, on one of the big crossbeams, it had taken the feather to its mud nest. There it would add it to the soft lining of feathers and grass that would cradle the eggs and later the young ones.

The woman called her husband and excitedly told him the story. She asked him to bring her feathers—lots of soft, fluffy feathers—from the chicken yard. Her husband knew that barn swallows seemed to prefer white feathers and so he gathered only white ones. When he returned with a small sack filled with them, the woman leaned from the open window and dropped one on the air. A breeze caught it, sped it upward and outward. A pair of swallows came racing from the barn; one caught the feather and turned deftly to carry it to the nest. The other bird, chippering excitedly, raced alongside its mate and both birds disappeared inside the wide-open barn doors.

Again the woman dropped a feather on the air and this time, two pairs of swallows came swiftly from the barn in a race to

### Hand-Taming with Nesting Materials

In June 1959, Mr. and Mrs. Carl A. Kennedy of Valencia, Pennsylvania, watched a pair of cedar waxwings start building a nest in a tree in their garden. To help them, Mrs. Kennedy cut pieces of pink and white cord about four inches long and walked into the garden with the cords on one hand. Soon one of the waxwings flew to her hand, picked up a cord, and flew with it to work it into the nest. After several trips, instead of alighting on her hand for cord, it perched on Mrs. Kennedy's head, reached down, and tried to pluck some of her hair for its nest. I've personally seen a tufted titmouse pluck hairs from the back of a live woodchuck sitting just outside its burrow, and another from the back of a gray squirrel stretched out on a tree branch, then carry the hairs to the nest.

catch the feather. Within a week, nine pairs of barn swallows were taking the feathers that the woman dropped out of the window for them each day. Several of the swallows became so tame that they snatched the feathers from her fingers as they flew past her window.

One morning she arrived at the window fifteen minutes later than usual. When she looked out she was astonished to see the barn swallows flying back and forth, calling in a way that she took to be a loud demand for more feathers! When she tossed one into the air, they all dived for it at once.

For three weeks the swallows came for feathers, until the end of May, when their nest-building seemed to be over. To the woman this had been a new experience with birds—one that had helped her regain her strength and had speeded her recovery. She vowed never to forget the swallows and said that, thereafter, they would never want for white feathers at nesting time and her barn doors would never be closed to them. They never were, as long as she lived.

**BARN SWALLOW**

| | |
|---|---|
| Number of Eggs Laid in Clutch: | 3-5 |
| Days to Hatch: | 14-16 |
| Days Young in Nest: | 21 |
| Number of Broods Each Year: | 2, sometimes 3 |
| Lifespan: | 6 to 8 years |

*Barn swallows range coast to coast, from Alaska and Canada south to Mexico. They nest in colonies in farm outbuildings, under eaves, bridges, and boat docks, in rock caves, holes, in natural cavities in cutbanks, or shelves built for them. The barn swallow nest is a cup of mud mixed with grass and feathers.*

# *Offering a Drink and a Bath*

any years ago Ernest Thompson Seton, a famous American naturalist, wrote an article, "Why Do Birds Bathe?" which was published in *Bird-Lore*, now called *Audubon Magazine*. Seton listed three kinds of baths that birds commonly take—sunbath, dust bath, and water bath. He told of having seen eagles, hawks, owls, grouse, quail, and buzzards taking sunbaths, of game birds and sparrows that took dust baths, and of other kinds of birds that liked a water bath. Of all the water-bathing birds he had seen, robins were the greatest bathers of all. One of them that he had watched soaked itself until it could barely fly.

But the question of *why* birds bathe he could not answer. He supposed that most birds did because it is healthful for them to do so, and he cited as an example the house sparrow. This is a vigorous and normally healthy bird that takes sunbaths, dust baths, and water baths, and will even wallow in the snow when no other baths are available.

As a result of Seton's article, many people wrote to him about the kinds of baths they had seen birds take, but not one could answer the question: "Why do birds bathe?" Dr. Arthur A. Allen, professor of Ornithology at Cornell University, wrote that dust baths usually help to kill bird lice and other insect parasites that sometimes get in a bird's feathers. These, by the way, are not the kinds that will infest human beings. As for other bathing, he could find "no biological reason for sun-baths or water-baths except as it seems to bring a pleasurable sensation to birds."

I have seen at least sixty-five different species of birds use the baths in our garden. After watching them year after year, I am convinced that they enjoy bathing quite as much as human beings do, particularly in warm weather.

Although the birds in our yard enjoy bathing, they come to our birdbaths far oftener for a drink. On hot summer days robins and other birds that move about on our lawn hold their wings out from

RULE OF THUMB

Clean your birdbaths and refill them with fresh water each day.

their sides and pant from the heat. These are the times when they drink and bathe most often. In July and August, we have had ten or twelve of several different kinds perched side by side on the edges of the baths, all drinking together. When many of them are drinking and bathing frequently, we clean the baths and refill them with fresh water each day. The strong flow of water from our garden hose does this quickly and effectively.

## THE BIRDBATH WITH THE SLOPING BOTTOM

The birdbath with a bottom that slopes downward from the outer edges toward the center, like the gradual descent of a sea beach or pond bottom, will attract the most birds. In it they can wade to the depth of water that suits them before they start bathing. They also like baths that have roughened bottoms that their feet can grip without danger of slipping. That is why concrete baths or those with a rock bottom seem to inspire them with greater confidence.

I made our first birdbath of concrete, at a cost of about one dollar for materials. I chose a sunny place on the edge of our lawn, next to a border of rhododendrons and other shrubs into which the birds could fly to preen and dry their feathers after bathing. There I scooped out a circular area about 3 feet in diameter. Along the outer edges of the circle I dug down only about 2 inches, but in the center I went down about 4 inches deep. Next I mixed with a hoe about 4 parts of sand to one part of cement, then added enough water to make a sticky, plastic mass, which I spread evenly in the hole to a thickness of about 1 inch. This made a cement-lined depression in the ground that sloped from ground level to about 3 inches deep at its center. It made an excellent, economical-to-build bath that lasted for a number of years. I could clean it out easily with a hose and broom, but it did fill up with leaves and other debris that the wind blew into it occasionally. Although it served its purpose satisfactorily, I came to prefer a birdbath that is above the ground—it is safer.

### The Shallow Birdbath

Birds will drink from and bathe in the same bath *(A)*. Follow the dimensions given in the text for making this birdbath *(B)*.

Birds will use the same water both for bathing and for drinking; but, in order to get the birds to bathe, the birdbath must be shallow. It should be no deeper than about 3 inches at its center and even less at its edges—perhaps no more than ½ inch to 1 inch deep where the birds enter the water. The deepest part of a bath of this kind will suit the larger birds—jays, grackles, and robins—and the shallows will be right for the smaller goldfinches, song sparrows, chickadees, and others. Small birds are afraid of deep water. You will notice that when they bathe along a stream's edge or in a woodland pool, they invariably hunt for places where the water trickles thinly over the rocks or the shallows close to the bank.

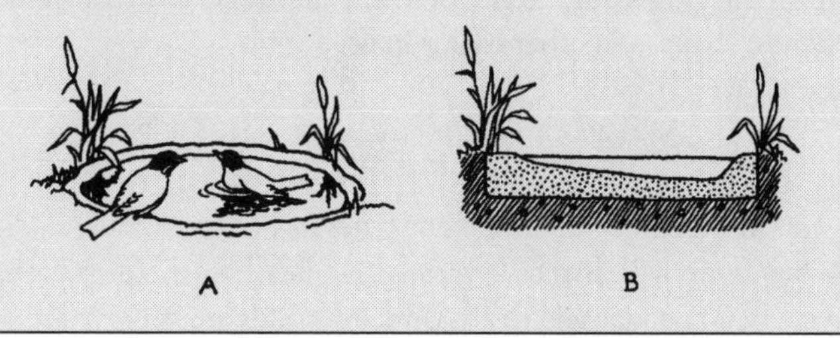

A B

RULE OF THUMB

Birdbaths with sloping bottoms attract the most birds.

## THE BIRDBATH ABOVE THE GROUND

One day, as I walked down our garden path, I saw a big gray cat move out of the shrubbery and bound toward our ground-level birdbath. I yelled, but I was too late to prevent the swift tragedy that followed. Even as I shouted, the cat leaped into the birdbath and struck down a robin that was too wet to fly quickly. With the robin in its jaws, the cat raced through our property-line hedge and disappeared behind a building in our neighbor's yard.

That was the only time that I had ever *seen* a cat catch a bird in the ground-level bath, but I could no longer let the birds use it. It was too close to the shrubbery for their safety. I could have

WARNING

Elevated birdbaths, which may be bought from bird equipment dealers and home improvement stores, are preferable to ground-level baths which can leave birds vulnerable to attack.

made a new cement ground bath in the center of the lawn where the birds could see a cat approaching and easily fly to safety, but I had a better plan.

A few days later I made a shallow concrete bath, 36 inches in diameter, and set it on a concrete pedestal and base that raised the bath 40 inches above the ground. I put the bath out in the open lawn, directly under the low-swinging branch of an apple tree into which the wet birds could fly at the first sign of danger. From their elevated position, they also could see any cat approaching long before it got within striking distance.

I especially recommend that you have an elevated birdbath in your garden, but it isn't necessary to build your own. Attractive ones, made of concrete or tile, may be bought from dealers in bird-attracting equipment, or from manufacturers of garden benches, chairs, and other backyard accessories.

## THE MAGIC OF FALLING WATER: THE DRIP BATH

TRY THIS

The steady fall of dripping water will make your birdbath even more attractive to birds.

You may have several different types of birdbaths in your yard, but none will attract a variety of birds like the drip bath, a device that you may add to any birdbath. Birds are strongly attracted by dripping water. So powerful is its lure that an occasional falling drop in your birdbath—perhaps only one or two a second—will attract warblers, flycatchers, some of the northern thrushes, and others that without falling water might never come to your bath. Many of these birds will not come to feeding stations. The drip bath is necessary if you would like to see as many species in your yard as it is possible to see in your particular region.

Although this is a simple arrangement, you may not like its appearance. I once hung a pail with a spigot near the bottom of it from a tree on our open lawn about 15 feet away from the birdbath. To the spigot I attached a 1/4-inch copper pipe and ran it downgrade, through a series of iron eye hooks that I sunk in the lawn. I ended the copper pipe at a point about 18 inches above the birdbath. I filled the pail with water, then turned the spigot on just enough that a drop of water issued from the end of the pipe and fell into the birdbath about every second. Although this worked, I found it unsatisfactory because it looked like a mad inventor's dream and did not add to the attractiveness of our yard. Later I moved the whole assembly of bucket and pipeline into our side-yard shrubbery and then shifted the elevated birdbath to the edge of the lawn near it. From spring until fall when leaves are on the trees and shrubs, they concealed the bucket and pipe from view, which made it quite satisfactory. In winter I stored the assembly in the cellar because it would freeze and fail to operate in cold weather. Also, its effectiveness would be wasted on the *wintering* birds in our Long Island backyard, which didn't need the drip of falling water to attract them.

For years the sharp *plink!* and *plop!* of falling water sounded out in our backyard. Altogether it attracted forty-five species of migrating birds that I am sure we wouldn't have seen without it.

**The Drip Bath**

One of the simplest drip-bath arrangements you can make requires only a 10- to 12-quart wooden bucket or metal pail. Drill a small hole in the bottom so that only an occasional drop of water will seep out. Suspend the bucket from a tree limb or some kind of artificial support no more than about 2 feet above the birdbath (any higher and the wind may blow the drops outside the bath). Some dealers in bird-attracting supplies now sell baths with drip attachments.

## The Birdbath in Winter

Like many of our friends who attract birds, every winter we have had the problem of keeping the water in at least one of our birdbaths "operating" during cold weather. To make the birdbath usable for birds, we had to thaw the ice with boiling water, sometimes three or four times a day on those days when the temperature dropped to 20 degrees above zero or lower. We never minded doing this. What worried us most were those occasional periods when we had to be away from home for a few days. Our hopper feeders filled to the brim with grain would feed our birds until we returned, but what about water?

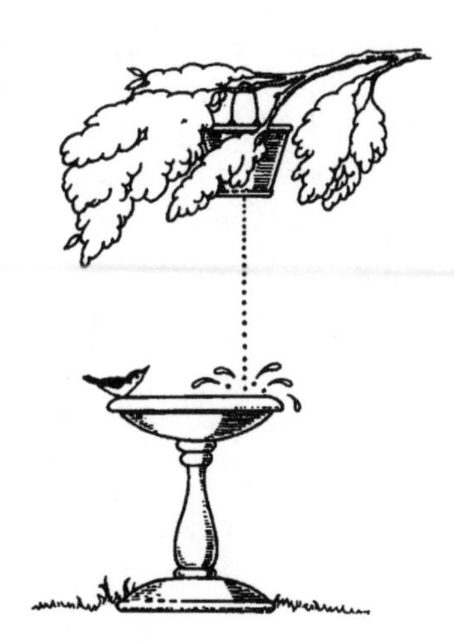

One winter that problem was solved very capably for us by a reader of *Audubon Magazine*. He introduced me to the "submersion" water heater for the birdbath. Here is what it is, and how it works:

Submersion, or waterproof, electrical heaters are usually made for fish aquariums in various watt sizes that will keep the water at a certain, even temperature. The warming power of these heaters will vary according to the wattage size of the heating unit. These are manufactured in 25-, 40-, 50-, 75-, and 100-watt heaters, which are sold by dealers in aquarium supplies and range from about fifteen to thirty dollars in cost. Heaters specifically for birdbaths are available from dealers in bird-attracting supplies and cost from thirty to fifty dollars. Heated birdbaths are also available. We used a 75-watt heater, which is probably ample for most birdbaths. The heating unit, sealed within a tube made of chrome, nickel, or Pyrex glass, was supplied with electric current by a short length of insulated, waterproof wire. For your birdbath, this can be plugged into whatever length of extension wire you need to reach from your

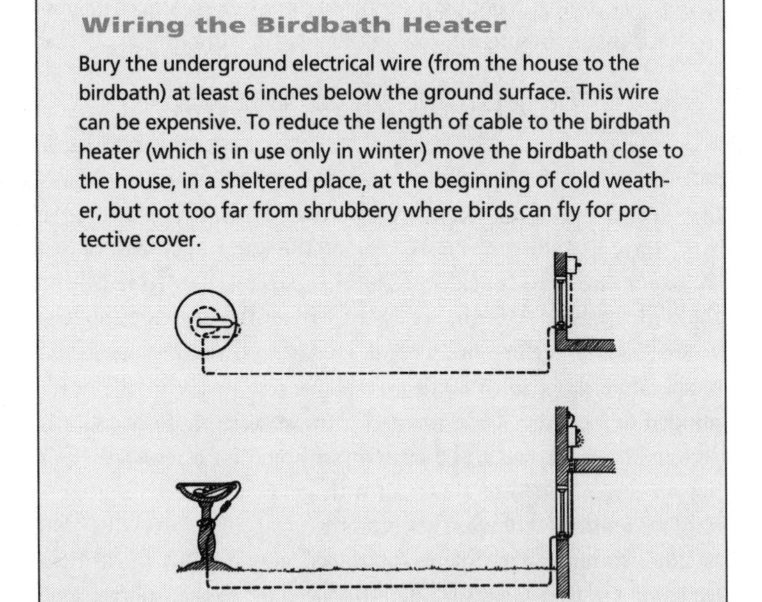

**Wiring the Birdbath Heater**

Bury the underground electrical wire (from the house to the birdbath) at least 6 inches below the ground surface. This wire can be expensive. To reduce the length of cable to the birdbath heater (which is in use only in winter) move the birdbath close to the house, in a sheltered place, at the beginning of cold weather, but not too far from shrubbery where birds can fly for protective cover.

house to the birdbath.

If your birdbath is some distance from your house, you might, as we did, run a durable weatherproof outdoor cable to it. Such cables cost about forty cents per foot and can be found at most do-it-yourself hardware centers. Back in the 1940s, when we first tried this, we used a lead-covered wire that we buried a few inches below the ground. Modern materials are superior to this. Make sure that you do not bury your wire where you may be digging in your garden at some future time. We ran ours from the house, under our lawn, to our north property line fence, then east, underground, close to the fence and beneath shrubbery all the way to the birdbath. Under shrubs and close to the fence there will be little danger that we will ever disturb or break the wire by digging, and there is not much chance of our doing so where it is buried under the lawn.

**TRY THIS**

An electric water heater will keep your birdbath operating during cold weather. Be sure to bury the wire under shrubbery or close to fences where there is little danger of accidentally digging it up later.

## CONNECTING THE BIRDBATH EXTENSION CORD AT YOUR HOUSE

We drilled a hole through the bottom rail of the sash of one of our basement windows (see sketch, page 116) and ran our lead-covered extension cord inside. We then called in an electrician to finish the job, because we didn't want to risk any wire work that might be a fire hazard. You can see from the sketch that the electrician installed a switch box near a base plug outlet inside our house, just below our living room window from which we watch the birds in our backyard. When the switch is on and the heater is operating, a red jewel light on the face of the switch box glows; when the switch is snapped off and the heater is not working, the light goes out.

Early in the morning during freezing weather we would switch on the heater, which soon thawed the ice in the birdbath. Our heater, turned on for only a few hours each day of the ordinary cold weather in our area, keeps the birdbath water thawed. If we have bitterly cold nights or days when the temperature drops near zero or below, we may keep the water heater operating all night and all day. The cost of operating the heater is so slight that since we have been using it, we have noticed little difference in our winter electric bills. In 1993, our local utility company said that our 75-watt birdbath heater, if it operated an average of 3 hours daily, for a 30-day month, would cost us only about 50 cents a month for electricity. The length of time that you need your water heater operating to keep your birdbath water thawed will depend on how cold or how mild the winters are where you live. You will soon learn, with a little experimenting, how frequently you will need to use it, and you will be more than repaid by the trips it will save you in cold weather, carrying hot water to thaw the ice in your birdbath.

**RUFOUS-SIDED TOWHEE**

| | |
|---|---|
| Number of Eggs Laid in Clutch: | 2-6 |
| Days to Hatch: | 12-13 |
| Days Young in Nest: | 10-12 |
| Number of Broods Each Year: | 2 |
| Lifespan: | 4 to 13 years |

*The rufous-sided towhee ranges across southern Canada and throughout most of the United States. It builds its nest of grasses, rootlets, twigs, leaves, and string on the ground under a bush or brush pile, also 1 to 5 feet up in vines, trees, and bushes, and lines it with horsehair and cattle hair.*

Our good friend, who introduced us to the easily installed and practical submersible water heater, now has what he believes is an even better idea. He has learned about a thermostatic heater on the market that can be set to maintain any desired water temperature. In winter his birdbath water could thus be kept at a temperature higher than 40 degrees by using this heater; it would use less electricity because the heater would go on automatically only when the water temperature in the birdbath dropped below the setting of 40 degrees. Such a thermostatically controlled aquarium water heater has a water-submersible heating unit sealed inside a metal tube 8 inches long and 1 1/4 inches in diameter, which is placed in the birdbath water. Attached to it is a 6-foot-long weatherproof electrical cord that can be plugged into an extension cord, which in turn can be plugged into an outlet, either inside or outside the house. This assembly can be arranged exactly like that of the submersible aquarium heater illustrated on page 116, and can be connected from the house to the birdbath in the same way. It differs from the fish aquarium heater in that it operates automatically. After the metal tube of the water warmer has been placed in the birdbath water, and the cord plugged into the house current, the warmer would go on automatically when the water temperature dropped to 45 degrees. When the water temperature in the birdbath rose to 50 degrees, it would shut off. Many aquarium stores carry such heaters, meant for fish tanks of 10 to 50 gallons.

## *The Strange Practice of "Anting"*

One October morning, some years ago, my wife and I were seated on a rustic bench overlooking our garden on Long Island, New York. We were watching three house sparrows taking their dust baths. They were lying flat on a sunny place in a flower bed with their body feathers slightly puffed out. Each had settled about a foot apart in its own small hollow, scratched out of the soil with the feet. They were shaking their wings and shifting their bodies about as they sifted dust through their feathers. Their eyes shone, they chirped, and every motion showed their pleasure.

The dust bath not only seems to suffocate body lice and other parasites in a bird's feathers but, according to one scientist's experiments, realigns the interlocking barbs of the body feathers and removes oil and moisture from the bird's plumage. Suddenly the sparrows got up, vigorously shook dust from their feathers, and flew off.

Then we saw a starling alight on our lawn. It seemed about to squat, and we assumed that it was going to sunbathe as robins and so many of our garden birds do on a hot day, sprawled flat, sometimes on their sides, gazing up at the sun. Through my binoculars, I could see the starling suddenly begin to pick up pale objects from the ground and dab them under its wings.

"That bird is anting!" I said excitedly. It was the first time in my thirty years of bird study that I had ever seen this strange phenomenon. The starling continued to pick up ants and to dab them under each wing. Then it flew away, and we walked quickly to the place where it had been standing. As I knelt on the grass, I smelled a familiar sweet odor, like that of the crushed leaves of our spicebushes in the garden border. In the starling's freshly dug holes, I saw several active golden-bodied ants, known to scientists as *Lasius*

**WOOD THRUSH**

| | |
|---|---|
| Number of Eggs Laid in Clutch: | 2-5 |
| Days to Hatch: | 13-14 |
| Days Young in Nest: | 12-13 |
| Number of Broods Each Year: | 2 |
| Lifespan: | 3 to 9 years |

*The wood thrush ranges the eastern United States, southern Canada, west to North Dakota, and south to Texas and northern Florida. Its nest is like a robin's, lined with rootlets, often with paper or cloth outside, but distinguished by dead leaves and mosses. The wood thrush nest is often found 6 to 50 feet up in a crotch or saddled on the branch of a tree, shrub, or sapling.*

*claviger*. The starling had disturbed them at the entrance to their underground gallery, and they had emitted their strong spicy odor in self-defense.

I found no dead ants on the ground, and I had not seen the starling eat them, although it may have crushed the ants in its bill before putting them in its feathers. One of the theories about this behavior is that birds benefit from the ants' defensive spray of formic acid.

Formic acid and other ant-fluid chemicals have certain insecticidal values when sprayed in the anting bird's feathers. These are thought to kill or repel bird lice, mites, and other annoying parasites that live on a bird's body or in its feathers. However, recent investigators believe that anting, usually practiced by birds during their summer and fall molt, is especially effective in soothing irritations of the bird's skin caused by its new feather growth.

At least 160 kinds of birds are known to practice anting and observations of it extend around the world. Besides ants, birds rub at least 40 kinds of substitute materials into their feathers, such as chokecherries, juice from the hulls of walnut shells, cigarette and cigar butts, hot chocolate, soapsuds, and sumac berries. Most of these materials are acid and may serve birds as effectively as the formic acid of ants.

In the spring of 1961, in a wooded tract of land in Chapel Hill, North Carolina, I watched a wood thrush hammering a land snail on the ground with its bill. Then it rapidly dabbed the snail under its wings, flank feathers, and tail. The thrush was "anting" with a snail—a never before reported event.

Some of the North American garden birds known to ant, besides starlings, are blue jays, catbirds, robins, tanagers, towhees, and cardinals.

# Part 2

## *Spending Time with the Songbirds in Your Yard and Garden*

# *Birdsongs and Sounds That Attract Birds*

y wife says that she has known for a long time why I spend hours in our garden, watching those fascinating details that make up a bird's life and a bird's world. She thinks I have more than my proper share of curiosity, like those romanticists who never see a hill in the distance without wanting to climb it to see what lies on its farther side. She is right, but what she didn't know, until I was able to prove it to her satisfaction one day, is that birds at times may be almost as curious about people as we are about birds.

One sunny morning in October a few years ago we were working in our yard, raking up russet and golden leaves from the grass under our tall white oak tree. We had sat down to rest on a bench on our terrace when my wife suddenly caught my sleeve. She pointed toward the birdbath.

"What is it?" she whispered.

A small bird, which had dropped down to our birdbath for a drink of water, had fluttered up into a large spruce tree in our next-door neighbor's yard.

I picked up the binoculars that we always keep close by when we are working in the garden. We have these at hand to watch any of the birds that are too far away for us to see them clearly with our unaided eyes. I looked at the spruce tree carefully through the powerful binoculars, but I couldn't see the bird, nor a movement in the tree that would tell us that it was still there.

We got up from the bench and walked toward the tree. We stopped a few yards from our property-line fence and looked up into the tree just beyond. Still we saw no bird. Not a sound came from the densely needled branches of the spruce.

I raised my right hand to my mouth, and pressed my lips against the back of my hand. By holding one corner of my mouth slightly away from my hand and by sucking air vigorously, I made a series of fine, high-pitched squeaks that sounded like a bird in distress.

**GRAY CATBIRD**

| | |
|---|---|
| Number of Eggs Laid in Clutch: | 2-6 |
| Days to Hatch: | 12-15 |
| Days Young in Nest: | 10-15 |
| Number of Broods Each Year: | Often 2 |
| Lifespan: | 4 to 10 years |

*Gray catbirds range throughout most of the United States except for the West Coast and the Southwest. The catbird nest, often found in garden shrubs 3 to 10 feet above the ground, is a ragged mass of sticks and weed stems that has an inner cup lined with pine needles and bark.*

Almost instantly a small bird fluttered out from the far side of the spruce tree, circled around toward us, and looped down to alight on the fence only twenty feet away. Focusing quickly on the bird, for it might stay there only a moment, I saw that it looked larger now, as all birds do when seen through a binocular that has a lot of magnifying power. The distance between us seemed shortened, too, as if it were only an arm's length away.

The little olive green bird shifted its position quickly, hopping about on the fence and turning its head from side to side. First it fixed its bright glance on us with one eye, then with the other. It seemed excited, and it puffed out its throat several times, as if it were calling, but I heard no sound.

That bird, for us, was a record-breaker. After studying it closely, I knew that it was an orange-crowned warbler, the *first* and the *only one* we had ever seen in our yard. This was a rare species in our area, and we wouldn't have seen it had not my wife discovered it. On the other hand, we wouldn't have identified it had I not used the "squeak," as ornithologists call it, to arouse the bird's curiosity and draw it into view.

## What Birds Are Attracted by Squeaking Sounds?

I can't remember who first taught me how to squeak to attract birds, but it must have been one of the older ornithologists of a group who took me on birding trips many years ago. Most ornithologists make this thin, high-pitched sound when they want to lure birds from tall grass, thickets, marshes, woodland undergrowth, and other places where birds are hidden from view. I have been using the squeak for years and find it particularly useful in late summer and fall.

In the spring it is much easier to see birds when the leaves on trees and shrubs aren't fully grown, and birds are conspicuous because they are singing almost constantly. But in late summer and autumn, they are inclined to be silent. This is the time of the year when strange birds, on their way south to winter in the tropics, can be in the shrubbery of your yard or neighborhood and you may never even suspect they are there.

In our garden, by squeaking on the back of my hand, I have brought close to us blue jays, chickadees, white-breasted nuthatches, catbirds, scarlet tanagers, brown thrashers, towhees, song sparrows, white-throated sparrows, red-eyed vireos, golden-crowned kinglets, and many other birds.

One summer day I made a series of squeaking notes near a catbird that nested in one of our shrubs. I knew that she was nesting there but had paid no attention to her at the time because I was trying to "squeak up" a small bird that I wanted to see, which had concealed itself in one of our yews. The catbird, flying from our vegetable garden with a cutworm in her beak to feed her youngsters, heard my squeaking sounds. Instantly she turned in the air and, with an enraged cry, flew down at my head. I ducked or I think she might have struck me. Then I stood looking up at her in astonishment as she perched on a limb overhead, scolding me.

The bird had never been aggressive toward me before and hadn't even protested when my wife and I had walked by within a few feet of her nest while she fed her family. But I had given a call—a squeak—that sounded to her, I suppose, suspiciously like that of one of her youngsters in distress. After that day, and up until her young ones left the nest, she scolded me whenever I came near. Obviously, she now looked on me as an enemy. I was not to be trusted.

**RULE OF THUMB**

When you squeak to birds, keep yourself at least partially hidden behind trees or shrubs.

## *Squeaking to Attract Birds*

For your squeaking to have its greatest effect upon birds—that is, to lure them closest to you—you should be at least partly hidden in a thicket of shrubs or under the down-sweeping branches of a tree where birds won't be able to see you as clearly as if you stood out in the open. Make yourself comfortable by sitting

**Whistling**

Some ornithologists don't use their hands at all to produce sounds to attract birds; instead, they make a combined whistling and hissing sound by using their lips, tongue, and teeth. This is most effective in attracting the small, golden-crowned kinglet and the brown creeper, probably because the hissing-whistling sounds resemble their own high-pitched, lisping call notes.

on the ground or on a stool or chair, and sit quietly without making motions. After squeaking a few times, pause to listen for answering birds, then repeat the squeaking to draw them as near to you as they will come.

Some birds—chickadees, titmice, and kinglets—will come to squeaking with remarkable boldness. Edward A. Preble, who was assistant editor of *Nature Magazine* and a former government biologist, had a tufted titmouse fly to his hand in response to his squeaking; it clung there, and pecked his finger quite sharply. W. E. Clyde Todd, curator of birds, Carnegie Museum, Pittsburgh, told me that on one of his scientific trips to Labrador he coaxed, by squeaking, a pair of boreal chickadees to come closer and closer to him. Finally they perched, side by side, on the barrel of the gun that he carried under one arm. Other birds—the shy thrushes, for example—won't come anywhere near you in response to your squeaking.

The curiosity in the birds—their emotional response that squeaking seems to arouse—varies, too. On some days all the squeak notes that I can utter do not draw one bird near, yet on the following day by squeaking I may easily attract swarms of small birds, especially migrating warblers in the fall of the year.

## MANUFACTURED BIRD CALLS, OR SQUEAKERS

I tried to teach my wife how to squeak birds, but she couldn't seem to master it. Besides, as she pointed out, she didn't like to wear her lipstick on the back of her hands or on her fingers. When she despaired over her inability to learn how to squeak, even *without* lipstick, I tried to comfort her by telling her of half a dozen American ornithologists I knew who had never been able to learn these methods either.

For these people, and for children too, there are manufactured bird calls, or bird callers, which are easy to use and are effective in attracting the attention of birds. The Audubon Bird Call, which is widely available in stores and catalogs, is a combination of a pewter plug and a rounded wooden receptacle. It is only two inches long and you can carry one easily in your pocket.

To make sounds with the Audubon Bird Call, you hold the wooden part of the caller in the fingers of your left hand. Then, with your right hand, you grasp the small round grip of the pewter plug between your thumb and index finger and turn it backward and forward as you would turn a key in a lock. This produces a squeaking or "creaking wheel" sound that arouses the curiosity of birds.

## Attracting Birds by Imitating Their Calls

Some birds can be attracted more readily by imitating their songs or call notes than by squeaking. Each spring I could tremendously excite the male northern oriole of a pair that nested in our yard simply by whistling a fair imitation of his song. This brought him down into a tree or shrub very close over my head, where he whistled at me in what I believed was defiance. Apparently he thought I was another male oriole that

### Calling the Birds to Dinner

Al Martin, who had remarkable success with hand-taming birds in Maine (see his *Hand-Taming Wild Birds,* Bond Wheelwright Co., 1963), believed in calling them to his feeders. When some pine grosbeaks he was hand-taming perched on his hands, he talked to them in the same tone he used to call birds at feeding time, but more softly—"Come and get it, come and get it." He found that by talking to his birds he got them to associate tones of his voice and certain words with food. This is an old training trick of falconers, some of whom whistle a repeated phrase to their birds while the birds are feeding. Thereafter, the falcon, while flying free, can be called in by the same whistle, loudly repeated, or by calls.

**BOBWHITE QUAIL**

| | |
|---|---|
| Number of Eggs Laid in Clutch: | 14-16 |
| Days to Hatch: | 23-24 |
| Days Young in Nest: See below | |
| Number of Broods Each Year: | 1 |
| Lifespan: | 7 to 10 years |

*The bobwhite quail ranges throughout the entire eastern United States, eastern Canada, west to Wyoming, and in the Southwest. It nests in a shallow depression on the ground (arched over and lined with grasses) in fencerows, at the edges of gardens, fields, woods, golf courses, and swamps. The young leave the nest with the parents at the time of hatching. The bobwhite quail will renest if an early nest is destroyed.*

had invaded his territory. He may have believed that if he came near me and kept whistling that he would drive away this strange unoriolelike creature that "sang" his oriole song.

Theodora Stanwell-Fletcher, author of *Driftwood Valley*, told me that she was more successful in attracting birds by imitating their calls or songs than by squeaking. She often started rose-breasted grosbeaks and fox sparrows singing, and she believed that she caused both hermit and wood thrushes to start singing by imitating their songs. She lived near a woodland in which screech owls and horned owls dwelt. By imitating their calls she brought them from the woods into the trees in her yard.

In autumn I entertained guests by taking them into our garden and imitating the whistled songs of the white-throated sparrows that spent the winter with us. When the white-throats first heard me, three or four of them fluttered from the ground up into our shrubbery and sat there whistling back to me their plaintive little song. This became a sort of game, which made me wonder if, perhaps, they weren't enjoying it too. As long as I stood there and whistled that sad little song, they would reply, even though it was late in the day and was growing dark.

## Calling a Covey of Quail

On one glorious afternoon in early autumn, I stood knee-deep in the grass of a hillside in western Pennsylvania. Below me a patchwork of farms spread the length of a valley that lay snuggled between two wooded ridges. All that summer I had heard the whistled call *bob-white! bob-white!* of quail that lived on the farms in the valley below and upon the slopes of the surrounding hills.

As I glanced down the hillside, over the yellow grass, I saw a quick movement. A brown form scuttled across an opening, followed by another, and then another, until I had counted ten plump, round-backed quail. Soft conversational notes drifted up

### Finger Squeak

I learned to squeak birds by sucking on the back of my hand, but an easy and equally effective method of attracting small birds is to use the back of one of your thumbs or the back of your index or middle finger. Press your lips on the top of one of the joints of your thumb or fingers and suck in air. This makes an easily produced whining squeak that sounds like the call of a young bird or the squeak of a mouse. In fact you can attract the small screech owl, at dusk or after dark, by making these squeaking notes, which sound like the mice on which these little owls feed. A few years ago one flew down in our yard and brushed against my hat as it passed. Apparently it had mistaken my squeaking for that of a mouse.

### Palm Squeak

Another method of squeaking is to press your lips against your open palm while holding the fingers of your hand straight to allow the sound to escape. Suck in as if kissing your open palm. This is easy to do and makes a louder but lower-pitched sound that is more attractive to the medium-sized birds—jays, flickers, robins, and others.

from them as they moved about. I decided to try an experiment.

Dropping quietly to the ground, I stretched out on my back, then raised slowly on my elbows high enough to see over the tawny grass tops. Softly I whistled *wurr-a-lee! wurr-a-lee! wurr-a-lee!*

All was silent for a moment, except a crow that called from across the valley. Then, faintly, I heard several notes that sounded like a reply. Once again I whistled *wurr-a-lee! wurr-a-lee! wurr-a-lee!*

For a moment all was silent again, then, clearly, hesitantly, the call came back to me like an echo out of the past. One of the quail in the flock had answered and its call took me swiftly back to the close of a day of hunting many years ago. A twelve-year-old boy then, I had accompanied a group of older men on a quail and rabbit hunt. As the purple shadows of dusk had fallen over the fields, and the guns had stopped roaring for the day, that pathetic call had come out of the darkness—the darkness of long ago. It was the rally call of the bob-white quail, by which the members of a covey that are not shot bring together at dusk those that are still alive. When they find one another, they draw up in a small, tight circle to roost on the ground for the night, each bird facing outward, ready to warn of danger in the darkness; each ready to fly for its life.

Softly I whistled again, and a reply, hesitant and sweet, came from so near that it startled me. I heard a slight rustling sound, and then several small brown birds ran out of the grass at my feet. Another suddenly appeared at my left side, not eighteen inches away! I dared not move my head, and I hardly dared to breathe for fear of frightening the bright-eyed birds that, like bantam chickens, were now gathering all about me. Without turning I could see them cocking their heads, first on one side and then on the other, looking up curiously at the long figure that lay before them in the grass.

Suddenly I realized, with almost a sense of shock, that they did not know me for a man—their enemy—but only as an inert mass from which a call had come that they understood. I felt ashamed for having fooled them, and I pitied the hesitancy in their actions, their puzzlement, and the struggle that seemed to be going on in them to understand.

Very faintly I whistled the rally call again, and a quail at my feet suddenly reached forward and thumped its bill smartly against the bottom of one of my boots. Its action was so comical, so much like a reproof, that I laughed out loud. In an instant, every bird turned and ran back into the grass.

Slowly I got up, and when I stood fully erect, some of the quail that hadn't run very far saw me. Perhaps they recognized me now as an enemy, for a group of them burst into the air with a thunderous roar of their short, rounded wings. Single, scattered birds, farther out in the grass, flew up in quick succession to join them. I watched them speed away until the last bird had disappeared far down the slope.

The memory of that experience still gives me a warm surge of pride to this day. I had discovered that I could speak the language of these trusting birds, and speak it so well that I could call them unerringly to my side.

# Hand-Feeding, or Hand-Taming

Years ago two men in West Virginia, father and son, both fine naturalists, shared a determination to hand-tame birds that led them to what we might consider today to be extraordinary lengths. They were the first to attract birds in their part of the country. The birds there were so shy that the naturalists felt it necessary to start feeding them by positioning their first feeder about 150 feet from the house, next to a shrub border where the birds usually remained under cover.

After the birds started eating from the feeder, the men moved it each day closer to the house until the birds had followed it to a window overlooking their yard. There the naturalists nailed a feeding board to the windowsill so that they could watch the birds close up from inside the house. This, they said, was about the limit

**BROWN TOWHEE**

| | |
|---|---|
| Number of Eggs Laid in Clutch: | 2-6 |
| Days to Hatch: | 11 |
| Days Young in Nest: | 8 |
| Number of Broods Each Year: | 2 to 3 |
| Lifespan: | 7 to 10 years |

*The brown towhee ranges throughout the western and southwestern United States, from Oregon to Texas. Its bulky nests are made of twigs, grasses, and plant stems, and are built either on the ground or up to 35 feet up (usually 3 to 12 feet up) in the densest part of a bush or tree.*

### Hand-Feeding Basics

There are a few simple rules for your first trials at hand-feeding. Do not try until you have had birds accustomed to coming to your feeders. If you have tried in mild weather without success, try early in the morning on a cold, storm-threatening day or in the morning after a snow- or ice storm when birds are extremely hungry. Be sure your feeders are empty before you try it.

If a bird flies to your palm to pick up a seed or crushed walnut kernel, do not stare into its eyes, as this may frighten it (some authorities skilled in hand-taming say you must not even swallow as this too may alarm the bird; but in my own intense preoccupation with a bird on my hand, I don't think I have ever done so). Keep motionless, and above all, do not make a quick movement with your hands, which birds seem to know can catch them. And never, *never* try to catch a bird that has learned to feed from your hands and trusts you. If you should, you may frighten it so badly that it may never come to you again.

### A Bird on the Lip

Edith Werner, who operated a tearoom years ago near Sebring, Florida, had remarkable success in training Florida scrub jays to come to her when she whistled a bright little tune. Within minutes they flew from all directions and alighted on her arms and shoulders and ate bread from her hands. These jays, which live in scattered tracts of sand, pines, saw palmettos, and shrub oaks, became so tame about the houses of winter visitors to Florida that they not only took peanuts from people's hands but from their lips.

to which most people go in attracting birds, but they wanted an even closer association with them. And so they clothed a stick with a sleeve of an old coat and attached a glove at the end and nailed the artificial arm and hand to the windowsill feeder. Withholding food from the feeder, they filled the artificial hand with black walnut kernels. This was a bird food that the West Virginia naturalists found to be "without a peer," favored by most of their wintering birds over all other foods.

Soon birds came to the walnut kernels in the gloved hand—tufted titmouse first, then chickadees and a Carolina wren. Now they tried their next experiment. They took away the stick and dummy gloved hand. One of the naturalists opened the window and put out his own gloved hand filled with walnut kernels. Quickly the same birds came, along with some downy woodpeckers, to take food from his hand.

Then the naturalists decided to go one step further with their taming of the birds. They built a dummy, or scarecrow, man and stood him alongside the main feeder in the yard, which they had emptied of food. They dressed him in a long dark coat, put a crushed felt hat on his head, a corncob pipe in his mouth, and a glove on his hand lying on the feeder. In the hollow of the crushed felt hat, in the bowl of the pipe, and in the dummy's gloved hand, they poured the irresistible walnut kernels.

Although nervous at first about this manlike figure, the birds finally perched on its hat, then on the corncob pipe and on the gloved hand, and picked up the walnuts without hesitation. A few days later, after the birds were coming fearlessly to the dummy, one of the naturalists took the dummy away and stood by the feeder. He had put walnut kernels in the top of the crushed hat he now wore and into the bowl of the corncob pipe clenched in his teeth. One of his hands, which he rested on the feeder, he had filled with walnut kernels. Quickly and confidently the birds came to him and fed—tufted titmice, white-breasted nuthatches, juncos, tree sparrows, and downy woodpeckers. Even white-throated sparrows and

## Some Wild Birds That Have Fed From People's Hands

Bluebird*
Bunting, indigo*
Catbird, gray*
Chickadee, black-capped
Chickadee, boreal
Chickadee, Carolina
Chickadee, chestnut-backed
Cowbird, brown-headed
Crossbill, red
Crossbill, white-winged
Finch, purple
Flicker, common (yellow-shafted )
Goldfinch, American
Grosbeak, evening
Grosbeak, pine
Hummingbird, black-chinned
Hummingbird, ruby-throated
Jay, blue
Jay, gray
Jay, scrub
Junco, dark-eyed (slate-colored)
Kinglet, ruby-crowned
Mockingbird
Nutcracker, Clark's
Nuthatch, red-breasted
Nuthatch, white-breasted
Redpoll, common
Redpoll, hoary
Robin, American*
Siskin, pine
Sparrow, chipping*
Sparrow, house
Sparrow, rufous-crowned
Sparrow, tree
Sparrow, white-crowned
Sparrow, white-throated
Tanager, summer*
Thrasher, brown ( cornbread )
Thrush, wood*
Titmouse, tufted
Towhee, rufous-sided (bread)
Vireo, red-eyed*
Vireo, solitary*
Warbler, pine
Waxwing, Bohemian (berries)
Waxwing, cedar**
Woodpecker, downy
Woodpecker, hairy
Woodpecker, red-bellied
Wren, Carolina
Wren, house

* Usually take food in summer from one's hands—suet, bread, caterpillars, earthworms, mealworms, raisins, currants, etc.—to feed young.

** Rather than for food, usually come to one's hands for nesting materials.

### The Consequences of a Bad Fright

Roy Ivor, a Canadian naturalist, had an experience that showed what a bad fright may do to an already tamed wild bird. In his large outdoor aviary, where his birds lived semiwild, a pair of catbirds had grown so tame from Ivor's everyday presence that they flew to meet him and perched on his hands. One morning, he discovered that a rat had gotten in his aviary and had killed some of his birds. To confine the catbirds while he got rid of the rat, Ivor had to net them and then, when taking them from the net, catch them in his hands. The catbirds were so frightened that it was two years before they would come to Ivor and perch on his hands again.

cardinals alighted on his hand and fed. And within touching distance came song sparrows, hairy and red-bellied woodpeckers, towhees, and even the usually shy Bachman's sparrow.

When the man moved, he moved very slowly, which did not seem to bother the birds. And all the while they fed, he spoke to them softly to get them accustomed to the sound of a human voice.

Now some of the chickadees came all the way to the kitchen door to be hand-fed. If the naturalists were slow to come outside with a handful of walnut kernels, the chickadees clung to the screens on windows and doors and hammered on them with their bills to attract their attention. If the birds did not see either of the men through the first-floor windows, they flew up to the second story and hammered on the window screens there.

One day, when a stranger on horseback came riding along the muddy public road past the farmhouse, one of the naturalists' conditioned chickadees dropped to the rider's hat. It remained there a moment, then flew up to a telephone wire overhead. Feeling the touch of the small bird, the rider stopped, looked about in surprise, then started on. Again the chickadee flew to the man's hat, apparently hoping that it held food. The stranger whirled about on his horse, jerked off his hat, and looked at it closely. Then he looked all about him. Nearby sat a tiny bird, but it seemed to offer no explanation, and so he rode on, shaking his head over the mystery.

After the elder naturalist died, his son left the farm, and the house was without an occupant for almost a year. With no one there to feed the birds, the naturalist had wondered and worried about how they would get along without their accustomed winter food. One spring day, on his return, he walked up the trail toward home. As he crossed a hill, in sight of the house, and walked through his old orchard, he heard a familiar *dee-dee!* When he drew a walnut kernel from his pocket, a chickadee alighted on his hand.

Today it is not difficult to get a wild bird to come to the hands for food. Perhaps it is easier than in the days of the West Virginia naturalists because so many people are feeding birds and hand-taming them. There is always the chance that a bird that comes quickly to your hands to feed—a black-capped chickadee especially—may already have been hand-tamed by someone else, although of this we can never be sure because of the general boldness of these small birds.

We followed the methods used by the West Virginia naturalists but did not use a dummy to condition birds to a human presence at the feeders. In the years we were in our suburban Long Island home, we virtually lived in our garden sanctuary, with its shrub border and an oak-beech woodland that edged a corner of our yard. In summer and often on mild days in winter, we sat reading, writing, or simply observing on chairs or on a chaise longue on the lawn near an open feeding tray set atop a post, with the birds feeding unconcernedly in it.

During our first winter there, early one cold morning, when I approached our empty center-of-the-lawn feeder, a chickadee flew to the pail of birdseed I carried and alighted on the rim. When I set the pail down on the snow, I scooped up a few sunflower seeds in my palm and rested the back of my hand on the empty open feeder. While I stood absolutely still, the chickadee made four quick flying trips to my hand, each time carrying away a sunflower seed to a

**WHITE-BREASTED NUTHATCH**

| | |
|---|---|
| Number of Eggs Laid in Clutch: | 5-8 |
| Days to Hatch: | 12 |
| Days Young in Nest: | 14 |
| Number of Broods Each Year: | 1 |
| Lifespan: | 5 to 10 years |

*The white-breasted nuthatch ranges throughout the United States and southern Canada. Its nest of wool, cow hair, bark, and feathers is often built high in the natural cavity of a large tree, in a woodpecker hole or knothole in oaks and elms. It also nests in birdhouses.*

limb of a nearby tree. There, while it held the seed firmly under its feet, it pounded the black-and-white-striped hull open with its bill and swallowed the exposed inner kernel. This was the first of a succession of chickadees and a few nuthatches that came regularly to our hands for food. All one winter, a very tame female downy woodpecker, at the slamming of our back door and our appearance in the garden, descended in little backward hitches down the trunk of an oak in our yard. There she clung to the bark about six feet above the ground and waited for us to approach. Then she leaned around the trunk to take delicate bites from a piece of suet held either in my wife's hand or mine, while we leaned motionless against the tree.

You should not be discouraged if you do not have such remarkable experiences with birds in the beginning such as those described here. Usually one must work a long time with individual birds to gain such trust. And after you have hand-tamed a bird or birds, always carry a few walnut kernels or sunflower seeds in your pockets while out of doors for their visits to your hands for food.

## *Inviting the Birds into Your House*

Of all the remarkable stories about wild birds in America accepting human company, I know of none that exceeds the experience of E. R. Davis of Leominster, Massachusetts. In October 1925, he saw a pine siskin at his window feeding shelf, and within a few days, 150 were feeding there on mixed birdseeds and sunflower seeds, along with some American goldfinches. That winter the siskins became very tame and within a short time regarded him as a friend. Whenever he stepped outside the door, down came the siskins from the trees to alight on his head, shoulders, and arms, looking for food. They had learned that he carried seeds concealed in a small box or in a dish to lure them to him.

When he invited them inside through a small sliding door in

### A Bird Nap

Although Ernest Harold Baynes never caught any of the birds that came so readily to him in New Hampshire, he could easily have done so with a chickadee that once ate so many walnut kernels while perched on his hand that it could eat no more. As it sat comfortably on his one hand, Baynes slowly cupped his free hand around the trusting bird until he could see only the top of its head. Possibly the chickadee became drowsy from too much food, from the warmth of Baynes's hands, or from reduced light because it closed its eyes, tucked its head into its back feathers, and went to sleep.

one of his kitchen windows, they came right into his home. Soon they perched on his head and shoulders or hopped about his desk where he was writing, looking for the handful of seeds they knew he would give them. On bitterly cold nights, some of them even spent the night in the kitchen, sleeping perched on a clothesline Davis had stretched across the room just below the ceiling.

In his upstairs bedroom, Davis had three windows that he left open at night regardless of the severity of the weather. Soon the siskins discovered his sleeping place and early each morning visited him, as Davis said, "to see that I did not oversleep." If he was asleep, one of the birds would arouse him by alighting on his face or side of his head and pulling at his hair to awaken him.

Davis decided to test the intelligence of these small birds by offering them a problem. One night, before he went to sleep, he placed a box of sunflower seeds on the windowsill near the head of his bed, with a glass cover over the seeds to prevent the siskins from getting them. Next morning, one of the birds came through the open window. It alighted by the box and looked through the glass cover but could not get at the seeds. Then it flew to Davis's pillow and pulled his hair to awaken him, but Davis pretended to be asleep. Again the bird flew to the windowsill, tried to get the

### The Reverend and the Scrub Jay

Some years ago, when I was editor of *Audubon Magazine,* the Reverend Benjamin Franklin Root sent me a photograph from California of a scrub jay on his hand eating a peanut. It even came into his car to feed when he called it.

seeds, and failing, returned again to Davis. This time it not only pulled his hair and pinched his ear but even tweaked his nose with its bill. Three times it tried to get the seeds and each time returned to Davis, who finally opened his eyes and reached for the box. Immediately the siskin flew to Davis's hand, waited until he opened the cover, then hopped into the box to enjoy its well-earned breakfast.

In another test of the problem-solving ability of the siskins, Davis arranged a feeding box, out of sight of the birds, behind a bedroom wall. Then he rigged a button on the front of the wall with a small platform just below the button on which the siskins could perch. If a siskin should press the button, a small quantity of seeds would pass down a chute to the bird's feet.

For a while, the siskins could not understand why Davis had put out a container that held no food. Then one discovered the button and pecked it sharply. Down came some seeds to the bird's feet. It ate them, then tapped the button again. Soon several of the flock had learned the trick by watching and responding to the behavior of the first one. Apparently this was an example, by the first siskin, of what students of behavior call trial and error learning. Usually it is the bolder, perhaps more investigative or more ingenious bird (or birds) that first discovers the answer to a new problem of how to get food, and the habit may be quickly picked up by others.

RULE OF THUMB

**Once a bolder or more curious bird has mastered a stunt like hand-feeding, other birds will quickly learn to follow its example.**

## *The Phonetics of Songs of Some American Birds*

Nothing has given me more pleasure than listening to the songs and calls of birds in my garden and then trying to reproduce the sounds using words and phrases of human speech. It is a good way for each of us to memorize the rhythm and pitch of the song phrases so that we can, by using words, recall the song. We imitate the sound of a ticking clock by saying *tick-tock*, and the bark of a dog by *bow-wow*, and we can imitate the sounds a bird makes in the same way.

Each of us should use our own words and phrases, but I must confess that I have borrowed many of mine from the phonetics of some of our older naturalists and ornithologists. I have found that most of them fit the birds' songs very closely.

TRY THIS

One of the best ways to memorize the rhythm and pitch of a particular bird's song is to practice imitating it using words and phrases from human speech.

Dr. Frank M. Chapman, dean of early American ornithologists, in 1900, following Thoreau's nineteenth-century words, thought the red-winged blackbird of our marshes, fields, and gardens sang a liquid *kong-quer-ree!* which, to me, in my North Carolina yard, sang *o-ka-lee!* John Burroughs thought the eastern bluebird called continuously *purity, purity;* in upstate New York and elsewhere, I heard it sing *her-er-bert* or *sher-er-bert*, which I think gives the soft, quavering quality of its voice.

Approving the use of words for birds' songs, British scientist Dr. W. H. Thorpe, a distinguished authority on birds' songs, and an associate wrote in a British ornithological journal:

> Many of us remember bird songs by means of little rhythmic strings of words, or a sentence, or perhaps by nonsense syllables. These are most effective in reminding us of what the bird says if we invent them ourselves. We have in mind such sentences as "Take two cows, Taffy" for the [British] wood pigeon . . . "I go to Guayaquil" for the tropical-American slate-colored grosbeak . . . or "Who are you?" for the pauraque . . .
>
> Of course, in fact, what the bird sings has phonetically little or no resemblance to these mnemonics of ours. The reason why they serve us so well is to remind us of particular bird songs chiefly because the rhythm is appropriate, and because the vowels that we include in such phrases reflect approximately the same changes in fundamental pitch as are found in the bird's song.

**Squeaking Mouse Toy**

If you have children you have undoubtedly seen the little rubber "squeaking" mice that they like, which are sold in many of the chain stores. I have used one of these and find it effective in exciting the curiosity of birds because of the sharp squeak you can make with it. Hold the rubber mouse in your palm, with its belly against your middle fingers and your thumb on its back. Press down quickly with your thumb, and release the pressure as quickly in a series of rapid motions.

Dr. Arthur A. Allen, whom I knew many years ago at the Cornell Laboratory of Ornithology, thought the indigo bunting sang *sweet, sweet, where, where, here, see it, see it,* an almost perfect rendition to my ears. In my North Carolina garden, I heard it sing one short song: *sweet-syrup, syrup, syrup, sweet!* Another, from a hedge outside my cottage door, sang a different song: *sip, sip, Ray—see-you-first!* and repeated it over and over for fifteen minutes, then silence.

There are probably more but similar renderings of the cardinal's loud, rich whistle than of the song of any other woods and garden bird. In 1900 Dr. Frank M. Chapman wrote that the song begins with long-drawn notes, *whee-you! whee-you!*, followed by a rapid *hurry, hurry, hurry, quick, quick, quick, quick.*

In North Carolina I heard a song, *cheer! more wheat, more wheat, more wheat!* In my mother's garden in southern New Jersey, one lovely March morning a cardinal sang *too-few! too-few! few! few! few! few!*

Thoreau fancied that the brown thrasher, singing its rich phrases, each twice over, from garden thickets, orchards, and country hedges, had advice for the Massachusetts farmer planting his corn—*drop it, drop it, cover it up, cover it up.* One of the most elaborate versions came in 1929 when a Mrs. H. P. Cook wrote to Dr. Chapman that a singing brown thrasher seemed to her like someone in a telephone conversation—*Hello, hello, yes, yes. Who is this? Well, well, well, I should say, I should say! How's that? I don't know! What did you say? What did you say?* and so on.

A. C. Bent, in one of his *Life Histories of North American Birds* bulletins, thought that all of these interpretations suggested the song—a succession of phrases of two to four syllables, each with a pause after it—loud, rich, and musical. A call note I often heard from a thrasher nesting in my North Carolina dooryard was a sharp *smack!* of alarm, sometimes followed by a whistled, melodic *pee-yore! pee-yore!*

Other birdsongs in my garden are much simpler—the sweet, soft whistled *spring com-ing!* or *fee-e-bee* of the black-capped chickadee, and its sputtering sharp call, *chicka-dee-dee-dee;* the tufted titmouse's *pee-to! pee-to!* or *pee-ter! pee-ter! pee-ter!;* the rufous-sided towhee's song, *Drink your tea-e-e-e*, and from my garden hedge, *sweet bird! s-i-n-g-g!* and its questioning call, *to-whee? chewink!* or *joree?* which has given the bird some of its common names.

The American robin, one of the most familiar of all of our garden birds, has had rendered many interpretations of its song, but none better than that of Mabel Osgood Wright, who, in her book *Birdcraft*, noted its song at dawn as *cheerily, cheerily, cheer up, cheer up!* which I think could hardly be improved on.

E. H. Forbush in New England wrote that robins' songs vary considerably, and one of them, to him, was "an imitation of the well-known phrase, '*kill 'em, cure 'em, give 'em physic!*'"

The red-eyed vireo, from the trees surrounding our garden in Connecticut, sang tirelessly in summer from morning to night (more than 22,000 songs in a day reported by a Canadian ornithologist). He has been called preacher bird by many old-time birders because his repetitious, emphatic phrases, with pauses between, are sermonlike, described as a conversational monologue: *You see it—you know it—do you hear me? Do you believe it?*

If you keep these words in mind, advised one authority, you will recognize the bird the first time you hear it sing.

The bobwhite quail, which often came into my North Carolina garden, is not a so-called songbird, but its songs and calls have great charm. The males whistle *bob-WHITE!* or *bob-bob-WHITE!* and *wheat-RIPE!* or, according to Forbush, *buck-wheat-RIPE!*

The assembly call of the covey, or flock, is a plaintive *ka-loi-kee, ka-loi-kee!* which, to Dr. Chapman, was *Where are you? Where are you?*

Margaret Millar, in her book *The Birds and the Beasts Were There*, thought that the California quail in her garden bid her to *sit-right-down, sit-right-down!* Ernest McGaffey, in 1922,

**BALTIMORE (NORTHERN) ORIOLE**

| | |
|---|---|
| Number of Eggs Laid in Clutch: | 4-6 |
| Days to Hatch: | 12-14 |
| Days Young in Nest: | 12-14 |
| Number of Broods Each Year: | 1 |
| Lifespan: | 6 to 14 years |

*The Baltimore oriole ranges throughout the eastern United States, west to the plains and prairies, and in southern and western Canada. Its nest is a gray swinging pouch, 5 to 6 inches deep, woven of plant fibers, hairs, and short pieces of colored yarn put out on lawn or shrubbery. The nest is usually suspended 25 to 30 feet up in the air from the drooping branch of an elm or maple.*

### The Back Door Dinner Bell

We never practiced calling our birds. In winter, they came without any bidding, at their main feeding times in early morning and in late afternoon. Often they also came for a short period to feed in midmorning and in midafternoon, and on cold, stormy days, they flew in a steady procession to and from the feeders. Quite possibly their visits were according to their periods of hunger or from associating the slamming of our back door with our visits to the feeders to fill them.

described the call of the Gambel's quail as *payt-eight-o, payt-eight-o, payt-eight-o*, the rhythm suggesting the calls of its close cousin, the California quail.

TRY THIS

It's a good idea to keep a record of your own phonetic renditions so you can recall them each time you hear a bird singing.

One of the most amusing interpretations of a bird's song came to me from Alexander Sprunt, Jr., at the time superintendent of southern sanctuaries for the National Audubon Society. One day in Florida we were driving along a street in Okeechobee. I heard a white-eyed vireo sing its emphatic song, described by Aretas Saunders not only as different from the songs of other vireos, but from all other birds. Alex drawled, "He says, '*Take me to the rai-l-l-r-o-a-d, QUICK!*'"

From a thicket by a small stream near my garden in North Carolina, I heard one sing *That's for real! chick!* Another seemed to say *Mister T! Take me there, quick!*

Thoreau wrote that the song of the song sparrow, whose singing in our gardens has endeared it to so many, has been rendered by old-country people as *maids! maids! maids! hang up your teakettle—ettle-ettle!* and *if, if, if-you-please-sir!*

I like Margaret Millar's version as she heard it in her California garden—*press, press, press, Presbyterians* (sometimes with one less press).

These are a few of my notes and those of others on the phonetics of birds' songs and calls. It might be fun in your own listening to see if the phonetics I have chosen are close to yours. Better, I suggest that you start keeping a record of your own phonetic renditions and apply them each time a bird sings. It will not only help in identifying the singer, but will make you much more attentive to birds' songs in general.

If you cannot identify the songs, and therefore the singers, of any birds—and many people cannot—the best and quickest way to learn them is to ask for help. Go into your garden with someone from your local bird club or Audubon Society who can hold up a hand at a song and say, "Listen, a catbird!" "There, a cardinal!" "That is a towhee!" or "Over there, a mockingbird!" and so on. Get to know their voices like you know those of your intimate friends over the telephone. Then you will not need to see the hidden singer in a tree or shrub in your garden to know who is out there.

TRY THIS

You might be able to enlist the aid of someone from your local bird club or Audubon Society who can easily identify the birdsongs heard most often in your garden.

Listening to recordings of birds' songs offers the advantage, in the absence of the live, singing birds, of hearing songs over and over, to learn them by repetition, and to enjoy them and the information given about each by the narrator. A good source is the Cornell Laboratory of Ornithology, Ithaca, New York, which lists available recordings in its descriptive catalogues of the Crow's Nest Book Shop. Many bookstores and record stores offer cassette tapes and compact disks that you can take into the field with you. The Borror Laboratory of Bioacoustics, Columbus, Ohio, supplies recordings to people interested in research on birds' songs.

## Are Mockingbirds the Best Singers?

In the early spring of 1938, I left a snowstorm in northern New York on my first trip to Florida. Mockingbirds were unknown then in that bitter winter land where I lived. For the first time I was coming into mockingbird country, and for the first time I heard their singing.

All along the highways and back roads of Virginia, the Carolinas, Alabama, and Georgia, I heard mockingbirds. Their singing—rich, strong musical phrases with great sweetness, in almost endless variations—brought to mind the lovely musical shuttle of the poet Whitman, repeated again and again, broken by pauses, a short silence, then on and on.

These were males that sang, declaring their guarded territories always from a prominent place such as roadside fence posts, house chimneys, rooftops, telephone wires, and treetops.

**TRY THIS**

You can learn to recognize and enjoy birds' songs by listening repeatedly to audio recordings, which are available at many book and record stores.

I stopped often to watch the gray-white birds—slender, long tails drooping, partly opened bills lifted to the sky, yellow eyes flashing in the sunlight, white throats pulsing with the trembling notes.

Everywhere along my trail in Florida I heard them—from the fragrant citrus groves of Miami to the blue-water highlands of Lakeland, across the wild Kissimmee prairies, in the sanctuary of the magnificent Bok Tower. At the old state capitol of Tallahassee, I heard them in the soft spring night, singing from moss-draped oaks and magnolia trees above columned plantation houses along cobbled streets.

A mockingbird singing in the moonlight of an old southern city—this was the setting that in romantic novels symbolized the antebellum days in the South. But it was not until 1961, when I went to live in Chapel Hill, North Carolina, that I could study the behavior of mockingbirds, to hear over and over their singing, and verify their astonishing ability to mimic the songs and calls of other kinds of birds.

**Patience Pays Off**

Dr. Frank M. Chapman, former head of the Bird Department of the American Museum of Natural History, at his feeding station in Florida, experimented with calling birds by whistling to them while they fed. It was several weeks before he got a response. Then one day, when only a catbird was in sight, Chapman whistled. Within less than a minute, cardinals, painted buntings, and many other of his birds came flocking to his feeder.

## COMPETING SINGERS

In January 1974, while preparing the article "Songs and Singing" for my *Encyclopedia of North American Birds*, I wrote to several ornithologists whose wide experiences with the singing of American birds I knew from personal association with them, asking for a list of the native American birds they considered to be our finest singers.

Dr. Arthur A. Allen, one of the founders and the first director of the Laboratory of Ornithology at Cornell University, wrote that he favored the hermit thrush. "I put him far ahead of the European nightingale which I probably have not heard at its best. My list would go something like this: (1) hermit thrush, (2) wood thrush, (3) veery, (4) mockingbird, (5) brown thrasher, (6) white-throated sparrow, (7) fox sparrow, (8) robin, (9) song sparrow, (10) rock wren. At least that is the way I feel this morning."

John Kieran, author of several fine bird books, wrote to me from his retirement in Massachusetts. He also preferred the hermit thrush. "I suppose," he wrote, "the mocker must be put near the top of the class *(Aves)* but I prefer the hermit thrush, especially when you hear it in the quiet of a spring or summer evening settling down over the New England hills as the sunset fades into twilight and the twilight fades into dusk. I'll take the hermit first, the mocker second, and the veery third. The best to me of the warbled songs is that of the rose-breasted grosbeak."

George "Doc" Sutton, from the University of Oklahoma zoology department, a bird artist, teacher, professional ornithologist, and writer of many distinguished books and articles, favored a third possibility. "My feelings about bird song are decidedly relative. I can't remember ever being thrilled by the mocker's song, even though I find it extremely interesting, trying to figure out what's being imitated. Similarly, I can't remember ever being thrilled by a hermit thrush's song, even though I've admired it greatly. But a

| MOCKINGBIRD | |
|---|---|
| Number of Eggs Laid in Clutch: | 4-5 |
| Days to Hatch: | 12-13 |
| Days Young in Nest: | 10-12 |
| Number of Broods Each Year: | 2 or 3 |
| Lifespan: | 4 to 12 years |

*The mockingbird ranges across the United States from northern California to the Atlantic coast and south to the Gulf of Mexico. It builds its nest of string, leaves, cotton, cloth, grasses, and twigs 1 to 50 feet above the ground in the fork of a small tree or bush, often in vines, cactus, red cedar, spruce, pine, boxwood, privet, and in country gardens and yards.*

robin's song, sung in the rain, has thrilled me over and over. There's something about the simple phrases that always reaches down deep inside me to the point of making me feel . . . well, worshipful, grateful for life, contrite for the things I've done wrong. I've had the same feelings on a bright windless morning out on the high plains when listening to a Cassin's sparrow's flight song. But I believe my favorite of them all is the white-crowned sparrow's."

## IS A BIRDSONG A WORK OF ART?

Nothing could have demonstrated the different and strictly personal choice of favored birds and their songs than these three letters. And that the character of each bird and especially the circumstances in which it is heard—so obvious in the Sutton letter—strongly influence the listener. How then could the relative musical qualities of different birds' songs be measured and rated, both subjectively and objectively?

Dr. Charles Hartshorne attempted it through years of study and evaluation, beginning in the 1950s. In 1973, his book, *Born to Sing*, surveyed and analyzed songs and singing of two hundred species with highly developed songs. James Fisher, a respected British ornithologist, believed that Hartshorne was the first mainstream scientist with the courage to suggest that birdsong might be considered a work of art:

> For years the bird-song analysts had been overwhelmed (and correctly—first things first) with the detection and diagnosis of the biological function of song . . . Now the musical, artistic superstructure is also conceded. Birds are, or some may be, musical poets, as the bird lovers from Aristotle . . . to Keats and Clare have been trying to tell us all along.

## THE TIME AND PLACE MATTERS MOST

I believe that birdsong is always personal and that the listener's emotional response is our judgment of a bird's singing in nature, *at the time and place* that we hear it—the singing of a veery from a darkening canyon at dusk; a piercingly sweet chorus of white-throated sparrows in the morning, floating up from an Adirondack spruce flat; the song of a field sparrow at the edge of an old fragrant field; the winter song of a fox sparrow at dusk from a thicket in a North Carolina swamp. I have never heard a living nightingale, but I am sure I have heard its equal.

I remember one such song that shall live always when I think of the music of birds. I had gone to bed early in a motel in Waycross, Georgia. The next day, April 20, 1954, I wrote in my journal:

> Last night, several times under the moonlit sky, I heard a mockingbird sing rapturously—a song of rolling, chopping notes and clear whistles that was like no mockingbird I had ever heard. This was an unusually fine singer, and out there, somewhere not far outside my motel window, he made the night even more exquisite with his lovely outpouring of song. It was incomparable and when I was awakened at dawn to a splendid chorus of birds' songs—of cardinals, Carolina wrens, and meadowlarks—I returned again in memory to my mockingbird of the night whose single glorious performance will follow me down the years.

# Hummingbirds and How to Attract Them

The professor of the summer nature study course at a university in Massachusetts looked around his classroom. He had finished his series of lectures about the life of the honeybee and fifty students had closed their notebooks on finished work. Freshly cut flowers, in vases on long laboratory tables, brightened the room.

Suddenly the professor raised his hand and pointed to the window. A tiny bird hovered in the air over the sill, its wings beating so rapidly that they were a blur on each side of its body. In the astonished silence that fell over the students, they heard the low hum of its wings as it poised in midair, looking into the room. In the bright sunlight its back glittered a resplendent green, and its throat feathers flashed a deep, glowing red.

The bird hesitated only for a moment, then darted into the room. With a sudden jerk, it stopped at a vase of flowers and, while it whirred its wings rapidly to keep aloft, it probed with its slender bill the hearts of each blossom. When it had finished with these, it sped around a group of students to another vase of flowers at the far end of the laboratory, turned quickly, and darted to another. There it searched several blossoms, gleaning from them such nectar and tiny insects as it could find, and in one swift movement flew out of the window and disappeared.

That happened in the summer of 1899. The uninvited but welcome bird-guest to that classroom of long ago was a male ruby-throated hummingbird, typical in its fearlessness of many of its kind. The beautiful little ruby-throat that comes to the flowers in our gardens will often come to feed from flowers held in the hands of people. Occasionally it flies fearlessly inside of houses, and has gently taken sugar from the lips of a man who knew the liking of hummingbirds for sweets.

**ALLEN'S HUMMINGBIRD**

| | |
|---|---|
| Number of Eggs Laid in Clutch: | 2 |
| Days to Hatch: | 15-17 |
| Days Young in Nest: | 22 |
| Number of Broods Each Year: | 2 |
| Lifespan: Possibly to 5 years | |

*The Allen's hummingbird ranges along the Pacific coast from Oregon to southern California. Its nest is a thick-walled cup, 2 inches across, made of dried weeds, leaves, plant down, and lichens and is found 1 to 90 feet above the ground, in the shade, saddled on weed stalks, in vines, shrubs, pine trees, and buildings.*

## THE FEARLESSNESS OF HUMMINGBIRDS

The fearlessness of hummingbirds is all the more remarkable when we learn that some of their kind are the smallest birds in the world. There are no hummingbirds in Europe or anywhere in the Old World. They live only in North, Central, and South America; about 330 different species, most of which are very small. A male ruby-throat, the hummingbird of the eastern United States, is about $3\frac{1}{2}$ inches long and weighs only 3 grams—$\frac{1}{10}$ of an ounce! Yet this Tom Thumb will attack crows, hawks, and eagles, which weigh from three hundred to sixteen hundred times its weight, if they fly anywhere near its chosen nesting territory.

I have never heard of a hummingbird attacking or chasing from its nest a human being, but one at a nest that I found gave me quite a start. The nest was on a limb of a beech tree in a woodland in western New York State. As I peeped into the nest to look at the two small white eggs that are no bigger than soup beans, the female suddenly appeared in front of my face, her wings droning like those of a large bee. (The female usually does all the work of building the nest, laying the eggs, incubating them, and raising the young ones.) As she hovered in the air a foot or two away from me, she appeared to be searching my face anxiously, as if to discover what my intentions toward her nest might be. About as big in diameter as a silver dollar, the nest was only about six feet above the ground, on a down-sloping branch of the tree. When I backed away from it a few feet, she settled down on her nest as calmly as if I did not exist and was brooding her eggs as I walked away.

## THE BIRD THAT REFUSED TO BE CHASED

The ruby-throat is one of at least nineteen kinds of hummingbirds from tropical America that spend some part of their lives in the United States. It is the only hummingbird that nests in our country east of the Mississippi River, and it nests both east and west of the Mississippi over a great area.

On a map of the United States, run your finger down the Atlantic coast from Ontario, Canada, south to Florida. Then trace a course from Florida west to Texas, then north through the Plains states to the Dakotas and Saskatchewan. Inside that area you have outlined, if you know the right places to look, you may find many nests of the ruby-throated hummingbird during its summer breeding season. After it has raised its family, it moves southward. Some ruby-throats winter in Florida and Louisiana, but most of them fly across five hundred miles of water in the Gulf of Mexico each fall to winter in Mexico and Central America.

Like other hummingbirds, the ruby-throat is especially pugnacious toward both its own kind and other birds during its nesting period. At this time it can usually intimidate most other birds, but not all of them. One spring morning near Hagerstown, Maryland, I saw a ruby-throat dart into the upper branches of a wild cherry tree. There it hovered threateningly in the face of a Blackburnian warbler, a bird with a beautiful orange-colored throat. The little warbler, scarcely larger than the ruby-throat itself, did not fly away from the threat of that sharp, slender bill pointed close to its face. Instead it lowered its head like a little fighting cock. Then it ruffled out its feathers aggressively, and followed with its own little bill every shifting movement of its tormentor. The hummingbird finally zipped away; the warbler lowered its feathers and went about its business of searching the leaves of the tree for small insects.

**ANNA'S HUMMINGBIRD**

| | |
|---|---|
| Number of Eggs Laid in Clutch: | 2 |
| Days to Hatch: | 16-19 |
| Days Young in Nest: | 18-21 |
| Number of Broods Each Year: | 2 |
| Lifespan: Possibly to 5 years | |

*The Anna's hummingbird ranges from northern California south to Baja California. Its tiny, lichen-covered, cup-shaped nest can be found in a bush or small tree, in semishade, 17 inches to 30 feet up, sometimes on the face of a cliff.*

# How Fast Can a Hummingbird Fly?

Both the flying speed of the hummingbirds and their superb maneuverability in flight give them an aerial advantage over many other birds. The little body of a ruby-throated hummingbird without its feathers is no larger than the end joint of one of your fingers, but the breast muscles that move the wings are enormous in proportion to the bird's size.

In 1934 a man driving an automobile along a highway leading out of Washington, D.C., saw a hummingbird flying alongside his car and moving in the same direction. He increased his speed to 50 miles an hour and the hummingbird stayed opposite him, as though it were challenging him to a race. Suddenly the bird put on a burst of speed and pulled steadily ahead of his car. As closely as the driver could gauge it, the bird appeared to be flying between 55 and 60 miles an hour. Another man living in Washington measured the speed of a ruby-throated hummingbird over a short flight of 53 feet in his yard. The bird traveled it in 3/5 of a second! This is a flying speed of about 60 miles an hour, which is far in excess of that of most songbirds and equals the cruising speed of some of the swiftest hawks. Apparently the airspeed and small size of hummingbirds make them safe from attacks by most birds of prey.

# Flower Colors That Attract Hummingbirds

If you like flowers and have a garden, you should find it easy to attract hummingbirds. Red, orange, and other bright-colored flowers, and especially funnel- or tube-shaped flowers, usually lure hummingbirds. The little ruby-throated hummingbird of our eastern states is so enamored of red that he has been known to hover in front of a man wearing a red necktie, over the head of a woman who wore a red ribbon in her hair, and has even inspected the red, sunburned noses of people!

A bird scientist who studied the flower color preferences of hummingbirds discovered that it is the *brilliance* of the colors of flowers that attracts them—they prefer intense colors to pale ones. Red, being the color complement of green, is the most conspicuous color that a flower can show. Orange-colored flowers, although not so brilliant as red ones, are more showy in deeply shaded swamps and woods, where hummingbirds are quick to find them.

RULE OF THUMB

**Hummingbirds are most attracted to brilliant red or orange funnel-shaped flowers.**

Green flowers are inconspicuous among green leaves, but in certain contrasting desert backgrounds and on the prairies during the dry season green flowers stand out. As far as I know, there are no green flowers in the eastern United States that attract hummingbirds. In the West the large green flowers of a tobacco plant, *Nicotiana paniculata*, attract them.

Hummingbirds also visit certain tubular-shaped flowers of blue, purple, and certain white ones that offer them nectar, small insects, and tiny spiders to feed on.

## *Flowers That Attract Hummingbirds in the East*

Near Cape May, New Jersey, trumpet vines, or trumpet creepers, grow wild over the hedgerows of the farms in that area. This vigorous-growing vine, which botanists and nurserymen call *Campsis radicans*, has a powerful attraction for hummingbirds when the plants are in bloom. Almost any summer day we can see half a dozen ruby-throated hummingbirds at one time by just sitting quietly near a flowering trumpet creeper. Sometimes we see clashes between hummingbirds when two or three of them try to probe the same orange-red flower at the same time. With angry squeaks and buzzing *z-z-z-z-z-z-z-t-t-t-t!* notes,

## Some Garden Flowers That Attract Hummingbirds

| ▶ Hummingbirds Known to Visit the Plants<br>▼ Garden Flowers That Attract Hummingbirds | ruby-throated hummingbird | black-chinned hummingbird | rufous hummingbird |
|---|---|---|---|
| *Abelia grandiflora*, glossy abelia | X | | |
| *Aesculus* (species), horse-chestnut or buckeyes | X | X | |
| *Agave americana*, century plant, also other species of *Agave* | | X | X |
| *Albizia julibrissin*, silk tree or mimosa | X | | |
| *Althea* (species), marsh-mallow, hollyhocks | X | X | X |
| *Anisacanthus thurberi*, desert honeysuckle | X | X | |
| *Aquilegia* (species), columbine | X | X | X |
| *Begonia* (species), begonia | X | | |
| *Beloperone californica*, chuparosa, and *Beloperone guttata*, shrimp plant | | X | |
| *Buddleia* (species), butterfly bushes | X | X | |
| *Campsis radicans*, trumpet creeper or trumpet vine | X | X | |
| *Canna generalis*, common garden canna | X | X | |
| *Caragana arborescens*, siberian pea tree | X | | |
| *Cestrum purpureum*, purple cestrum | X | | |
| *Citrus* (species), orange tree | X | X | X |
| *Cleome spinosa*, spider flower or spider plant | X | | |
| *Crataegus* (species), hawthorn tree or thorn-apple | X | | |
| *Delonix regia*, royal poinciana | X | | |
| *Delphinium* (species), larkspur | X | X | X |
| *Dianthus* (species), pinks, sweet Williams | X | | |
| *Erythina cristi-galli*, cockspur, coral-tree | | | |
| *Eucalyptus* (species), eucalyptus trees, especially the scarlet-flowering kinds | | | |

| Costa's hummingbird | broad-tailed hummingbird | blue-throated hummingbird | Anna's hummingbird | Rivoli's hummingbird | calliope hummingbird |
|---|---|---|---|---|---|
| | | | | | |
| | | | | | |
| X | X | X | | X | |
| | | | | | |
| | | | | | |
| | X | | | | |
| | | X | | X | X |
| | | | | | |
| | | | | | |
| | | | | | |
| | | | | | |
| | | | | | |
| | | | | | |
| | | | | | |
| | | | X | | X |
| | | | | | |
| | | | X | | X |
| | | | | | |
| | X | | | | |
| | | | | | |
| | | | X | | |
| | | | X | | |
| | | | | | |

## Some Garden Flowers That Attract Hummingbirds

| ▶ Hummingbirds Known to Visit the Plants<br>▼ Garden Flowers That Attract Hummingbirds | ruby-throated hummingbird | black-chinned hummingbird | rufous hummingbird |
|---|---|---|---|
| *Fouquieria splendens*, ocotillo, coach-whip, or vine-cactus | | X | |
| *Fuchsia* (species), fuchsia | X | | |
| *Gilia* (species), scarlet gilia, sky-rocket, skunkweed, birds-eyes | X | | X |
| *Gladiolus* (species), gladiolus | X | X | |
| *Grevillea robusta* and *Grevillea thelemanniana*, silk-oak | | | X |
| *Hamelia erecta*, scarlet bush | X | | |
| *Hemerocallis* (species), day lilies | X | | |
| *Heuchera sanguinea*, coral bells | X | | |
| *Hibiscus* (species), shrub althea, rose mallow, Confederate rose, rose-of-China, crimson eye | X | X | |
| *Impatiens balsamina*, garden balsam | X | X | |
| *Ipomoea* (species), morning glories | X | X | |
| *Iris* (species), iris | | X | |
| *Jasminum* (species), jasmine or jessamine | X | X | |
| *Kolkwitzia amabilis*, beauty-bush | X | | |
| *Lantana camara*, lantana and other species | X | X | X |
| *Lilium* (species), lilies | X | | |
| *Lonicera japonica*, japanese honeysuckle, and *Lonicera sempervirens*, trumpet honeysuckle | X | X | |
| *Lupinus* (species), lupines | X | | X |
| *Malvaviscus drummondi*, Texas mallow, or Spanish apple | X | | |
| *Melaleuca* (species), bottlebrushes | | | |
| *Monarda* (species), bee-balm, horse-mint, Oswego tea, bergamot | X | | |

| | Costa's hummingbird | broad-tailed hummingbird | blue-throated hummingbird | Anna's hummingbird | Rivoli's hummingbird | calliope hummingbird |
|---|---|---|---|---|---|---|
| | | x | | | | |
| | | | | | | |
| | | x | x | | | |
| | | | | | | |
| | | | | x | | x |
| | | | | | | |
| | | | | | | |
| | | | | | | |
| | | | | | | |
| | | | | | | |
| | | | | | | |
| | | | | | | |
| | | | | | | |
| | | | | | | |
| | | | | | | |
| | | | | | | |
| | | x | x | | x | |
| | | x | | | | |
| | | | | | | |
| | | | | x | | |
| | | x | | | | |

## Some Garden Flowers That Attract Hummingbirds

| ▶ Hummingbirds Known to Visit the Plants<br>▼ Garden Flowers That Attract Hummingbirds | ruby-throated hummingbird | black-chinned hummingbird | rufous hummingbird |
|---|---|---|---|
| *Nepeta cataria*, catnip | X | X | |
| *Nicotiania glauca*, tree tobacco | | X | |
| *Nicotiania sanderae*, flowering tobacco | X | | |
| *Parkinsonia microphylla*, palo verde | | X | |
| *Pedicularis* (species), wood betony or lousewort | | | |
| *Pelargonium* (species), geraniums | X | | |
| *Penstemon* (species), beard-tongue | | X | X |
| *Petunia* (species), petunias | X | X | X |
| *Phlox* (species), phlox | X | | |
| *Poinciana gilliesi*, bird-of-paradise | | X | |
| *Quamoclit* (species), star ipomoea, cypress vine, cathedral climber | X | | |
| *Rhododendron* (species), azaleas | X | | |
| *Robinia pseudoacacia*, black locust | X | | |
| *Salvia* (species), sage | X | | X |
| *Saponaria officinalis*, bouncing Bet or soapwort | X | | |
| *Scabiosa* (species) scabious, mourning bride, or pin-cushion | X | | |
| *Tecomaria capensis*, Cape honeysuckle | X | | |
| *Tritonia* (species) montbretia, red-hot poker, kniphofia | X | | |
| *Tropaeolum majus*, garden nasturtium | X | X | |
| *Verbena* (species), verbena | X | X | X |
| *Vinca major*, periwinkle | | X | |
| *Vitex agnus-castus*, chaste-tree | X | | |
| *Weigela* (species), weigela or weigelia | X | | |
| *Yucca* (species), Adams-needle, Spanish bayonet, or Spanish dagger | | X | X |

| Costa's hummingbird | broad-tailed hummingbird | blue-throated hummingbird | Anna's hummingbird | Rivoli's hummingbird | calliope hummingbird |
|---|---|---|---|---|---|
| | | | | | |
| | | | X | | |
| | | | | | |
| | | | | | |
| | X | | | | |
| | | | | | |
| | X | | | | |
| | | | | | |
| | | | | | |
| | | | | | |
| | | | | | |
| | | | | | |
| | | | | | |
| X | X | | | | |
| | | | | | |
| | | | | | |
| | | | | | |
| | | | | | |
| | X | | | | |
| | | | | | |
| | | | | | |
| | | | | | |
| | | | | | |
| | | | | | |

they dart about with amazing swiftness. Usually they chase each other so rapidly that the birds to us are just a streak as they zip in and out of the hedgerow trees and shrubs.

In our Long Island garden, hummingbirds that arrived early in spring first visited the scarlet flowers of our Japan flowering quince. They also liked the orange and red flowers of our columbines and the brilliant red early-blooming azaleas. As the season progressed they came to our tatarian bush-honeysuckle (which had small, yellowish white flowers), to our morning glories, early larkspurs, weigela bushes, nasturtiums, Japanese honeysuckle vine, tiger lilies, bee balm *(Monarda)*, and scarlet sage.

You will also attract hummingbirds if you plant gladioluses, cannas, petunias, hollyhocks, geraniums, lilies, coral bells *(Heuchera)*, cardinal climbers, scabiosa, and cleome, or spider flower. All of these are favorite hummingbird flowers. You can buy the seeds of them at nurseries, and sow them yourself, or you may wish to buy the plants and set them out in your garden.

RULE OF THUMB

Plant your flower bed or garden with the tallest flowers in back, the medium-height flowers in the middle rows, and the shortest at the front.

When planting your flower bed, put the plants that will grow tallest in the back, the medium-height flowers in the middle rows, those that will be shortest in the front of the bed. Try also to plant each kind of flower in clusters, or solid groups. This will make your flower bed more showy and more effective when the plants are in bloom.

Hummingbirds also like to visit the red, trumpet-shaped flowers of the coral, or trumpet, honeysuckle, *Lonicera sempervirens*. This is an attractive vine that grows wild from Florida west to Texas, north to the New England states, and west to Iowa and Nebraska. Hummers like the purple flowers of buddleia, or butterfly bush, the pink and white flowers of horse chestnut, or buckeye, trees, and the yellow flowers of *Caragana arborescens*, commonly known as Siberian pea tree. A more extensive list of U.S. plants that attract hummingbirds appears on pages 154-159.

## THE MYSTERY OF THE NIGHT-FLYING "HUMMINGBIRD"

I remember, as clearly as though I had seen it yesterday, the honeysuckle vine that covered the east side of the porch of my boyhood home in the country. Hummingbirds often came there to sip nectar from the white honeysuckle flowers in the daytime, but at dusk another creature came that for a long while puzzled me. On summer evenings, just before the moon came up, I sat on our front steps waiting for the whippoorwills to call, and watching for the mysterious stranger. While I sat there, I breathed in the sweet fragrance of the honeysuckle and listened to the night voices of crickets and katydids.

Suddenly—silently—the stranger came. In the pale light of the rising moon a tiny creature, whose wings moved so swiftly that they were a blur, hung before a white honeysuckle flower. Hovering, backing up, moving rapidly from one blossom to another, it probed each flower with what seemed, in the dim dusk, to be its long, slender beak. I had always seen hummingbirds flying about in the daytime. Could this be a hummingbird that flew at night?

One evening I waited by the side of the honeysuckle vine until the mysterious dusk-flyer came close. In one swift motion, I swung my cap and swept it out of the air. I trembled excitedly, for I was sure that I had captured some strange and unknown creature that I had never seen before. Carefully I opened my cap until I could pick it up in my fingers. I remember my amazement to this day at what I saw, for I had caught a moth! So strikingly did it resemble a hummingbird that if I hadn't caught it, I might have gone on believing that hummingbirds flew in the dark.

Later a local naturalist told me that I had caught a hummingbird moth, that it was one of many of its kind, some of which fly about by day, others by night. What I had thought was its beak was its long, harmless tongue, with which it sips nectar. My mystery

**RING-NECKED PHEASANT**

| Number of Eggs Laid in Clutch: | 10-12 |
| --- | --- |
| Days to Hatch: | 23-25 |
| Days Young in Nest: See below | |
| Number of Broods Each Year: | 1 |
| Lifespan: | 8 years |

*The ring-necked pheasant ranges from southern Canada across the northern United States, from New England to Oregon and California. It usually nests on the ground, in a depression lined with bits of grasses and weeds, in fields of grass or grain, and sometimes at the edge of a garden, hedge, or roadside ditch. The young leave the nest with their mother after hatching but before they are able to fly.*

had been solved and I had held in my hands the only creature that you or I might reasonably mistake for a hummingbird.

## *Some Hummingbird Flowers of the South*

If you live in the southeastern United States, you will find that ruby-throated hummingbirds will visit the pink, fragrant flowers of your mimosa, or silk tree. Botanists and nurserymen call this tropical plant, *Albizia julibrissin*. It grows thirty or forty feet tall, and is hardy as far north as Washington, D.C. Hardier varieties of it can survive the winters north to Pennsylvania and New York, and in some of the New England states.

The red buckeye, *Aesculus pavia*, a small tree up to twenty feet tall, is another southern plant whose dark red or purplish flowers lure hummingbirds. It grows from Virginia south to Florida and Louisiana. The trumpet creeper, *Campsis radicans*, so much favored by hummingbirds in the northeastern states, grows south to Florida and Texas.

In southern Florida, the scarlet bush, *Hamelia erecta*, and the royal poinciana, or flame tree, *Delonix regia*, are noted especially for their attractiveness to hummingbirds. Others—the scarlet rose mallow, *Hibiscus coccineus*, which grows wild in the swamps of Georgia and Florida, the purple cestrum, *Cestrum purpureum*, common lantana, *Lantana camara*, fuchsias, cannas, butterfly bushes, and jasmines—are excellent "hummingbird plants."

### A HUMMINGBIRD EXPERIMENT

In 1889, Caroline G. Soule of Brookline, Massachusetts, developed an idea for feeding hummingbirds that may have been the first experiment of its kind. Her idea was so good and so practical that people who attract hummingbirds to their gardens have used the same method to this day.

For a week she had watched a ruby-throated hummingbird visit the red flowers of the trumpet creeper vine that grew luxuriantly over a dead tree in her backyard. Sometimes when nectar in the flowers may have been low, the hummingbird became so eager for the sweets that, instead of thrusting its head and slender bill inside the flower tube, it slashed open the petals at their bases and drank nectar while it hovered outside the flower.

How could the woman help the bird? Then it occurred to her: *Why not offer it a supplemental supply of liquid sweets?* But how could she offer food to the bird so that it would accept it?

One day she sketched a trumpet creeper flower on a piece of stiff paper that she had fashioned into the tubular shape of one of these flowers. Next, she painted the outside of the paper a brilliant orange-red to match the color of the living flowers. Inside the paper flower she set a small open-mouthed vial and then wired her artificial flower in a natural position among a cluster of trumpet creeper flowers. Using a mixture of about one part of sugar to two of water, she filled the little bottle within the paper flower and stepped back a few yards to watch.

Soon a hummingbird came to the trumpet creeper. Deliberately it hovered before one flower cluster after another until it had circled almost completely about the vine. With its next move, if it followed its circling course, it would arrive before the cluster of flowers within which the woman had wired the artificial flower.

Would the hummer fail to be attracted by the paper flower with its sugar water, and pass it by, or would it stop to drink? Its humming wings had carried the little bird before the artificial flower. It hung before it in midair, then unhesitatingly dipped its bill into the sugar water. Again and again it drank, left the sugar water, and then returned, as if it could not quite believe that it had discovered such a rich food supply. Thereafter it preferred the sugar-water mixture to the nectar in the living flowers, and the woman had to fill the vial twice a day to keep the bird supplied.

**CALIFORNIA QUAIL**

| | |
|---|---|
| Number of Eggs Laid in Clutch: | 12-16 |
| Days to Hatch: | 21-23 |
| Days Young in Nest: | See below |
| Number of Broods Each Year: | 1, sometimes 2 |
| Lifespan: | 6 to 10 years |

*The California quail resides along the Pacific coast from southern Oregon to Baja California, and also in Nevada, Utah, and other western states. It nests in a slight hollow lined with grasses and leaves, in the ground under a bush, hedge, or in clumps of grass or weeds in an orchard, vineyard, cactus, or under a brush pile, especially near a house or garden or beside a much-traveled road or path. The young leave the nest with parents at the time of hatching.*

One day the woman took the artificial flower with its sugar-filled vial off the trumpet creeper and stood nearby, holding it in her hand. When the hummingbird came to the vine it suddenly veered away and came fearlessly to the paper flower in her hand. There it hovered and drank its fill from the little bottle!

In that gesture of confidence the hummingbird assured the woman that her experiment has been an undreamed of, and probably a heretofore unheard of, success.

### Large Fishes and Frogs

One fall day a family in Santa Barbara, California, were seated by a lotus pool in their yard. A hummingbird flew to the pool and hovered momentarily a few inches above the water. Perhaps an insect struggling on the surface may have aroused the bird's curiosity. Whatever the cause, the slight distraction that held it over the pool brought the bird's life to a quick end. With a loud splash a bass broke through the surface of the water, flashed into the air, and swallowed the hummer. Large frogs in ponds and along stream banks also catch hummingbirds, perhaps when hummingbirds hover too near them in the same way.

## WHAT TO DO WHEN FLOWERS WON'T ATTRACT HUMMINGBIRDS

For thirty-five years a man and his wife had attracted birds to their estate in New Hampshire. They had operated feeding stations, put up birdhouses, and kept their birdbaths filled with fresh, clean water. They had planted on their property trees, shrubs, and vines that bore fruits that birds like and that offered them protective cover and nesting places. By their devotion to attracting birds and their commonsense methods of doing so, this New Hampshire couple had lured to their property almost every kind of bird that migrated through or nested in their region. They had attracted all

### Hummingbirds and Windows

In Massachusetts a hummingbird dashed against the picture window of a home and broke its neck. The bird had apparently seen the reflected image of trees and grass in the large window. This had created the illusion that they were part of the yard and it had flown into the glass at great speed. Another tried to fly through two windows that were opposite each other in a room. Apparently the bird thought that it could fly through them to the lawn and trees on the other side of the house. The windows were closed and the bird was killed when it struck the glass. See page 240 for a way to prevent birds from flying into windows.

of the locally common summer birds, *except one*.

Even though they had planted many kinds of flowers, they had been unable to attract hummingbirds to their garden.

In the summer of 1928, the man's wife read an article by a woman in Maine who had attracted hummingbirds to small bottles about 2 inches long, covered with bright-colored ribbons. The bottles, or vials, she had filled with sugar water as a food offering to the hummingbirds. The New Hampshire woman and her husband immediately followed this woman's example. The results were remarkable.

Eight years later, in 1936, a visiting bird scientist stood on the sunny veranda of this New Hampshire home. While he watched, twenty hummingbirds fed at one time from various small bottles filled with sugar and water that were fastened to the veranda railing and to twigs of shrubbery planted nearby. Within an hour he counted fifty hummingbird guests that came to the feeders.

## *What to Feed Hummingbirds*

At first the New Hampshire couple had offered the hummingbirds a mixture of strained honey and water, but this fermented easily and became unpalatable. (Later, Dr. Augusto Ruschi, a Brazilian expert on hummingbirds, both captive and wild, and author of the great 2-volume work *Aves do Brasil*, discovered that honey, when fermented, produces a fungus that affects the tongues of hummingbirds and eventually kills them. He does not recommend honey as a food for hummingbirds, even if the honey has been centrifuged, because as soon as the bottle is opened it can become contaminated and immediately infect the food solution and hummingbird food containers.) The New Hampshire couple turned to sugar water as an offering to their hummingbirds; and even song sparrows, purple finches, orioles, chickadees, nuthatches, and downy and hairy woodpeckers came

WARNING

**Honey becomes contaminated with a fungus that affects the tongues of hummingbirds and eventually kills them.**

WARNING

Mix sugar with at least 8 parts water when you offer it to hummingbirds.

to drink from the vials. However, sugar water must not be offered hummingbirds in too rich a mixture as it may cause enlargement of the liver with harmful effects. Dr. Ruschi recommends sugar water as the only basic food to feed hummingbirds in the garden and he recommends 25 grams of cane sugar dissolved in 200 grams of water, or about 1 part sugar to 8 of water.

In the beginning the woman had covered the outside of the glass vials with differently colored ribbons, until she noticed that the hummingbirds emptied those vials first that were wrapped in *red*. After that she used red ribbon.

## HOW MUCH WILL HUMMINGBIRDS EAT?

In 1936 she had thirty vials fastened at intervals all along the veranda. Some of them she had attached to twigs that she had clipped from shrubs. She had fastened these twigs with their attached vials to window frames and to the porch railing. Other vials she suspended on fine wires hung from an overhanging trellis. Each day the hummingbirds and other songbirds that came to drink the syrup emptied the vials three or four times. Between May 9 of that year and September 14—about four months—the birds ate 65 pounds of sugar.

A woman in Pennsylvania, who attracted hummingbirds to sugar water, used 2-ounce bottles with large mouths and no necks, which she covered with red satin. She wired each vial at a 45-degree angle to the tops of 36-inch stakes, which she "planted" upright in her flower beds. Some of the vials she attached to shorter stakes, which she put in flower pots on a backyard terrace. Hummingbirds came to the eight bottles of sugar water regularly and emptied them several times a day. From May 1 to about August 1 they ate 10 pounds of sugar.

In a fascinating feeding experiment in Iowa, a woman bird scientist fed wild ruby-throated hummingbirds in her backyard for seven consecutive summers. She discovered that each bird ate a

### Spider Web Death Trap

One of the most dramatic death traps for hummingbirds is built by the spider—a creature that ironically also provides many hummingbirds with one of their most important nesting materials.

I once watched a ruby-throated hummingbird start to build her nest. First she built a foundation for it of cottonlike down that she gathered from fern stems and oak leaves. Using lichens, bits of moss, and small pieces of bark, which she took from the trunks of trees, she built up the sides of the nest. I did not count the number of trips she made to bring strands of spider webs to her nest, but she used them frequently to glue her delicate materials together. It is risky for the little hummers to gather spider silk from the strong webs of the larger spiders because they sometimes get caught.

To gather spider silk from one of these big, vertically built, wheel-shaped webs, a hummingbird must fly to within a few inches of it. There, suspended on its rapidly whirring wings, it faces the web and plucks strands of it loose with its bill. This brings the bird dangerously close to the coarse, sticky web. One slight error in the bird's judgment—a swing too near—and those whirring wings are enmeshed, and the bird is helpless.

Ruby-throated hummingbirds in the East, and Costa's and Anna's hummingbirds in California, caught in spiders' webs, have died of exhaustion and starvation before they were discovered. People have found others in webs while the birds were still alive, and have released them unharmed. There must be many other hummingbirds caught in webs each year that we never hear about.

**EASTERN KINGBIRD**

| | |
|---|---|
| Number of Eggs Laid in Clutch: | 3-5 |
| Days to Hatch: | 13-16 |
| Days Young in Nest: | 13-14 |
| Number of Broods Each Year: | 1 |
| Lifespan: | 3 years |

*The eastern kingbird ranges across Canada to Nova Scotia, south to northeastern California, and east to Texas and Florida. This bird usually nests well out on the limb of an isolated tree or in low shrubs, but it also nests on the rain gutters of houses. Its bulky nest is built of twigs, straw, weed stems, and feathers, 2 to 60 feet above the ground.*

level teaspoonful of sugar each day. This is about $1^1/2$ times the average weight of each. If a man weighing 200 pounds ate sugar at the same rate, he would eat 300 pounds of sugar each day!

These days, there's no need to make your own feeder unless you're the sort that enjoys such projects; hummingbird feeders are available at many hardware stores, as well as through mail order from dealers in bird-attracting equipment such as those listed in the Appendix.

## *Hummingbirds of the West—the Largest and Smallest in the United States*

The Rivoli's hummingbird, about 5 to 5½ inches long, is generally considered the largest hummingbird in North America. It does not come into the United States very far north of Mexico. As far as I know, it nests only in Arizona and New Mexico, high in the mountains near the Mexican border.

The ruby-throated hummingbird of the eastern states is small, but the calliope hummingbird of the West, which is only about 2¾ inches long, is the smallest hummingbird in our country. If you live anywhere within that region from southern California north to Oregon and Washington, eastward to Montana, Wyoming, and Utah, and south to Baja (Lower) California, you may find this tiny bird nesting, or possibly migrating through your area in the spring. You are not likely to see it in the fall because it migrates southward through the Rocky Mountains region, instead of moving along the Pacific coast slope, where it migrated northward in spring.

The calliope likes to feed at the flowers of paintbrush (*Castilleja*), columbines, wild roses, orange trees, wild gooseberries, and hawthorns.

Besides the calliope and the ruby-throated hummingbird of the East, six other hummingbirds—the rufous, black-chinned, broad-tailed, Allen's, Anna's, and Costa's—push northward well into the United States in spring. The rufous breeds the farthest north—to Alaska—and is the most widely distributed and probably the most abundant hummingbird in the West.

The broad-tailed, a hummingbird that usually nests in the

### Avoiding Ants

To keep ants out of hummingbird feeders, suspend the vials by thin wire from the top of a window frame or from the underside of the eaves of the porch or the house, in front of a window where you can observe them.

John V. Dennis, in his book *The Complete Guide to Bird Feeding* (Knopf, 1976), notes that an effective way of keeping ants from crawling along wires or other supports for hummingbird feeders is to coat the wires with salad oil; he also coats the feeders to keep bees and wasps from settling on them.

Another technique I have heard about is to pierce a metal can with a hook, so that the eye is inside the can and the hook protrudes beneath. Seal it with epoxy and fill the can with water. Then hang the can from a wire attatched to the eye of the hook inside, and hang your feeder on the hook. Ants don't like swimming from the wire to the can's edge, and they'll leave the suspended feeder alone. Such hook-and-can devices can also be purchased from dealers in bird-attracting supplies.

mountains, is the hummingbird of the Rocky Mountain region. It nests in Montana, Idaho, Wyoming, Utah, Colorado, New Mexico, and southwestern Texas, and westward to Arizona, Nevada, and eastern California.

The broad-tailed follows along with the blooming period of its favorite plants, moving gradually up the mountain slopes to its nesting areas in spring, and down them in fall on the way to its winter home in Guatemala in Central America. In the San Francisco Mountains of Arizona its principal food plant is the beautiful scarlet trumpet flower, *Penstemon barbatus torreyi*. After this flower finishes blooming, the broad-tailed visits the flowering beds of a blue larkspur, *Delphinium scopulorum*. This wildflower grows high in the mountains, from southern California to Alaska. The broad-tailed is also attracted to different kinds of sage (or salvia), scarlet paintbrush (*Castilleja parviflora*), nasturtiums,

### Hummingbirds and Wildflowers

*Castilleja*, the Indian paintbrushes, or painted cups, are parasitic on the roots of other plants and depend upon them in part for their lives. That is why people who have tried to transplant paintbrushes to their gardens have been unable to keep these plants living. If you dig up paintbrush in a clump of sod, with some of its associated plants, you may be able to keep it alive in your garden for a little while, but it is far better to let paintbrushes grow where nature has planted them. There hummingbirds will have the use of them, and these flowers will continue to brighten the countryside.

lupines, penstemons, and ocotillo, which botanists call *Fouquieria splendens*, a spiny desert shrub with scarlet flowers. Gilia, agave, and trumpet honeysuckle are others of its favorite flowering plants.

## How Fate May Direct Our Interest Toward Birds

I remember the instinctive reaction of a little girl the first time that I showed her a hummingbird. The tiny creature, on droning wings, hovered before a flaming-red canna flower. The child turned to me, her eyes wide with mixed wonder and fear. "But it's so *little!*" she whispered. "How does it live?"

That is the way I feel about a hummingbird every time that I see one. To me, one of the marvels of this natural world is that a creature so small, and seemingly so fragile, can stay alive from day to day. Hummingbirds have great flying speed, sufficient to make them safe from most birds or other animals that might eat them—if they could catch them. But there are other threats to their lives. Some of them are so subtle, or innocent-appearing, that we wouldn't think of them unless we had our attention called to them.

These are tragedies that our sympathy for hummingbirds makes us want to prevent. Other than covering picture windows with mesh (see page 240), we can do little to prevent hummingbird accidents. We must remember that most of these "accidents" arise from natural hazards that are a part of the hummingbird's world. They serve to skim off only a part of the hummingbird population, and are Nature's way of preventing the plight of "The Old Woman in the Shoe," who, you will remember, "had so many children that she didn't know what to do."

It is strange how destiny sometimes directs our interest toward birds, and how that interest may benefit not only ourselves, but the birds too. A number of years ago I heard of a businessman who was forced to move to Arizona because of ill health. He was not a bird

### Dispelling Bees and Wasps

A friend at Chapel Hill, North Carolina, when much troubled by bees and wasps at her hummingbird feeders, moved her wire-suspended feeders out of the sun, where the insects seemed the most active, and hung them in the shade. Then, when insects gathered at the feeders, she turned on her vertically rotating lawn sprinkler, positioned so that the water spray, with each rotation, struck her hummingbird feeders. The water dispelled bees and wasps and helped wash away the sticky fluid adhering to the outside of the feeders. The spray did not bother the hummingbirds, which not only continued to come to the feeders but seemed to enjoy bathing in the fine water spray by flying back and forth through it.

scientist; but, when he settled in Phoenix, he almost immediately became interested in the black-chinned hummingbird, which nests in and around that city and throughout much of the Southwest. For four years he studied these birds, which are closely related to the ruby-throated hummingbird. His discoveries about their nesting and feeding behavior and social habits are interesting, and he has also given us specific information about attracting black-chinned hummingbirds to gardens.

For example, he discovered that the black-chinned, like the rufous, calliope, Costa's, and broad-tailed hummingbirds, which migrate long distances into the United States, seem to time their arrival with the blooming of certain plants. In spring, when the black-chins arrive in Phoenix after spending their winters in Mexico, some flowers at which they feed are already in bloom. One of these, the American aloe, or century plant, *Agave americana*, has white flowers that bloom on a tall stalk, and give off a scent like that of butterscotch caramel. These are an excellent source of nectar for several kinds of hummingbirds. Another, the tree tobacco, *Nicotiana glauca*, which grows up to twenty feet tall, is a native of South America. It has been naturalized in Texas, California, and in other states and hummingbirds like to visit its yellow, tubular-shaped flowers for their rich supply of nectar.

Others that the black-chins favor are the red, orange, or yellow flowers of lantana, and the blossoms of orange trees, of other citrus trees, and of the shrimp plant, *Beloperone guttata*. The black-chins also like the flowers of the paloverde, or littleleaf horse bean, *Cercidium microphyllum*, ocotillo, yuccas, morning glories, gladioluses, nasturtiums, hollyhocks, jasmines, and butterfly bushes *(Buddleia)*.

**EASTERN WOOD PEWEE**

| | |
|---|---|
| Number of Eggs Laid in Clutch: | 2-4 |
| Days to Hatch: | 12-13 |
| Days Young in Nest: | 15-18 |
| Number of Broods Each Year: | 1 |
| Lifespan: | 7 years |

*The eastern wood pewee ranges from southern Canada south to Florida and west to the Dakotas and Texas. Its nest (which looks very much like a knot atop a branch) is a thick-walled cup, 3 inches in diameter, built of plant fibers and weed stems on a horizontal limb, 15 to 50 feet up in an oak, maple, locust, or elm.*

## HOW TO ATTRACT BLACK-CHINNED HUMMINGBIRDS

During his first summer in Phoenix, only one black-chinned hummingbird came to the man's yard to feed at the few

**Hummingbird Water Sports**

In California, Emerson Stoner reported that when he watered his garden a female Anna's hummingbird often darted through the stream of the hose. One day, she discovered that she could ride this solid stream of water. She alighted on it crosswise, as she would alight on a twig, and allowed herself to be carried forward. She kept repeating her newly learned trick in what seemed to Stoner a delightful spirit of play.

canna flowers that a former tenant had planted. The man wanted to attract more of them in order to study these birds. The following spring he added beds of nasturtiums and larkspurs to the cannas. A female black-chinned hummingbird then came to his yard and nested. He also put up hummingbird feeders—small bottles filled with sugar water that he attached to shrubbery or to sticks "planted" among the flower beds. Within a short time, five female black-chinned hummers came regularly to feed from them each day. By late summer at least eight others and a pair of southward migrating rufous hummingbirds were visiting his sugar-water feeding stations.

In the spring of the following year, two female black-chins nested in his yard, and six fed regularly from the feeders. Others, which had not discovered the honey and the sugar-water mixture, visited the flowers in his yard each day.

Hummingbirds are quick to defend their nesting territories against the intrusion of other birds. With so many hummingbirds coming to a yard where two of them were already nesting in bushes on opposite sides of the house, tension soon built up. It was to touch off some small explosions of hummingbird tempers.

## The Right of "Private" Domain

That second spring, the first of the female black-chinned hummingbirds to build a nest in the yard had driven off all hummers except a handsome male that she had accepted as her mate. Daily the man watched her grow more aggressive as her instincts to defend her nesting territory grew with her urge to build her nest and lay eggs. She drank frequently of sugar water, and she guarded the feeders from other birds by watching over them from her perch on the twig of an ash tree in the front yard. From this point she could see all the feeders and chase away most birds, even before they got to the feeders.

To the man, it must have been obvious that the little bird considered these feeders and the yard about them to be hers! Any other bird that came near trespassed upon her private property! Her persistence in guarding the feeders was so effective that the man determined to outwit her. He did want other hummingbirds to enjoy his hospitality, too.

One morning he moved one feeder to the rear of the house, and another to the front of the house. He did not move the third one, which remained alongside the house near a row of oleander bushes. The aggressive little hummer could now see only the front-yard feeder from her perch. And while she drank sugar water from any one feeder, she could not see a bird drinking from either one of the others.

It seemed like a peaceful settlement, a division of the hummingbird feeders in which all birds could share. The trick worked—for a few days. Then the little female hummingbird, whose intelligence the man had underestimated, caught on.

Instead of keeping her guard post on a twig of a tree in the front yard, she suddenly shifted to a eucalyptus tree from which she could watch both the feeder at the rear of the house and the one in the side yard! By shifting between the tree perches, she could now watch all three feeders and continue her dominance over the birds in the yard.

## THE QUEEN'S DEFEAT

Day by day, the little female grew more domineering over her chosen territory. Now she chased away not only hummingbirds but house sparrows, which previously she had ignored.

Eventually, the reign of all tyrants, even small, harmless ones like hummingbirds, must end, for nowhere is the dominance of one creature over another more likely to shift than in the world of birds.

One day a strange female black-chinned hummingbird arrived in the yard. She had visited the flower beds, and flew to one of the hummingbird feeders. The little female, from her perch in the eucalyptus tree, saw the stranger for the first time. Like a

**BLACK-CHINNED HUMMINGBIRD**

| | |
|---|---|
| Number of Eggs Laid in Clutch: | 2 |
| Days to Hatch: | 16 |
| Days Young in Nest: | 20-21 |
| Number of Broods Each Year: | 2 or 3 |
| Lifespan: | 5 years |

*The black-chinned hummingbird ranges throughout southern Canada and down the Pacific coast, east to the Rockies, and south to Mexico. Its nest, a cup of plant down 1 1/2 inches across and 1 inch high, is saddled on a drooping branch or fork of a limb 4 to 8 feet up, often over water or a dry creek bed, and also in gardens.*

tiny thunderbolt she hurled herself downward at the bird in a blazingly swift dive.

The newcomer, hovering at the feeder, seemed in danger of being knocked out of the air by that fiery assault. At the last moment, she shifted to one side and the attacking female buzzed by, stabbing at the air. Again and again the little female dived at the stranger. Each time the newcomer changed her position ever so slightly, avoiding the attacks with a calmness that must have been maddening to the bird so determined to drive her away. After each unsuccessful dive the little female returned to her perch in the eucalyptus tree. There she fluffed out her feathers and chattered angrily at the stranger, but the unwelcome visitor continued to feed about the yard and to ignore her.

At last the female that had dominated the yard so successfully had met more than her match. As the stranger moved confidently about the yard from flower to flower, and from feeder to feeder, she showed that she intended to stay. Gradually, the little female's attacks upon her grew less and less frequent, until they ceased altogether.

**Maintenance of Hummingbird Feeders**

At our Little Neck garden, we used simple vials to hold our sugar water and were very particular to keep them clean. Every day or two, we emptied any sugar-water solution left in the vials, washed each inside and out with warm water and a small stiff brush, then refilled them with a fresh supply of sugar water.

## THE NEW "YARD BOSS"

The little female now had little time to guard her territory. She had built her nest in one of the oleander bushes in the yard, had laid her eggs, and was busy incubating them. Her aggressive attacks on other birds had almost ended, but another feathered volcano had erupted in her place. The stranger, coming into her own breeding cycle, became another tyrant and drove away all other hummingbirds, just as the little female had. Between these two, however, there now seemed to be peace, for the man never saw any further skirmishes between them.

The newcomer mated with a male black-chinned hummingbird that came to the yard each day. She built her nest in a bush in a corner of the yard farthest from where the little female had hatched her family. Each evening at dusk, the man saw the newcomer's mate come to the bush where she nested. There he roosted on a twig not far from where she sat brooding her eggs.

The individual lives of birds, to those of us who study them, are filled with many vague, half-formed pictures that we are not always able to fill in. Even while we watch over them, things happen to birds for which we can give no accounting—no logical explanation. One day the newcomer disappeared and she never returned. Only her nest and its two small white eggs remained, to remind the man of the swiftly changing fortunes of creatures that must take their chances in the natural world.

The little female, as far as the man knew, had a more fortunate summer. She continued to return to the feeders. With her now came two young hummers which, by her tolerance of them, the man took to be the youngsters she had raised in the oleander that spring.

### A Dragonfly and a Hummingbird

In Ontario a man and his mother were walking through a woodland when they heard a peculiar rattling noise among the dead leaves on the ground. When they walked to the source of the sound, they found a hummingbird lying on the ground with a large dragonfly clinging to its back. The dragonfly had seized the little bird by the neck and had pinned it to the ground. The man chased the dragonfly, picked up the dazed bird, and held it quietly in his hand. In a few minutes the hummingbird recovered and flew away.

## *Other Birds That Like Sugar Water*

Besides hummingbirds, some fifty or sixty other kinds of birds, such as orioles, catbirds, brown thrashers, mockingbirds, and the red-bellied woodpecker, are fond of sugar water. One year, while I was editing *Audubon Magazine*, Ruth Thomas of Morrilton, Arkansas, wrote an article for us in our bird-attracting column about attracting orchard orioles to sugar water. Orchard orioles are fond of flower nectar, which they get by sipping it from the hearts of blossoms in their winter home in the tropics.

Mrs. Thomas discovered that at least seven pairs of these birds, which nested in her area, often visited the hummingbird feeding vials. In alighting on the resting perches of these feeders, they jolted them so severely that they often splashed half the sugar water out of the feeders. She solved this problem and got

**RED-BELLIED WOODPECKER**

| | |
|---|---|
| Number of Eggs Laid in Clutch: | 3-8 |
| Days to Hatch: | 11-12 |
| Days Young in Nest: | 24-26 |
| Number of Broods Each Year: | 1, or 2 to 3 in the South |
| Lifespan: | 7 to 13 years |

*The red-bellied woodpecker ranges from Minnesota east to Ontario, south to central Texas, and east to the Gulf coast and Florida. Its nest is usually dug in an old stump, in utility poles, fence posts, or in the decayed top, stub, or limb of a dead tree, with an entrance hole 2 inches in diameter, 5 to 70 (but usually less than 40) feet up. It may also nest in an old woodpecker hole.*

### More About Sugar Water

If you are just beginning to attract hummingbirds, it is best to use the rich solution of one part of sugar to two parts of water because hummingbirds prefer the sweeter mixtures. After mixing the sugar and water, boil it for several minutes. This will help to slow fermentation after the syrup is put outside in the feeder. After filling the feeders, the extra quantity may be safely stored in your refrigerator.

Once the hummingbirds are coming to your feeders, you can dilute the solution to one part sugar to four parts water and eventually to the one part sugar to eight parts water recommended by Dr. Ruschi to protect hummingbirds against the possibility of getting enlarged livers from the richer solutions. Hummingbirds can detect small differences in the sugar-water formulas, and you may discover as others have that they will not drink a solution with less sugar than one part to eight parts of water.

the orioles away from preempting the hummingbird feeders by dressing a small cold cream jar in a red ribbon frill, necessary only in the beginning, to give it a semblance to a flower. She wedged the cup securely into a wooden frame nailed to a corner post of her garden. The orioles immediately accepted the cup and in late June, when the young fledged orioles also discovered it, she set up an additional one to take care of her extra orioles and other birds that now found the sugar water attractive. Each spring, the migratory orioles returned to her garden around April 12. Mrs. Thomas had the cup on her corner post filled with sugar water, and the orioles flew straight to it.

# *Planning Your Yard and Garden with Birds in Mind*

Planting your yard to beautify it and to attract birds is a lot like building a house. You start by planting your biggest, sturdiest plants—the shade trees and evergreens—which are the foundations. Then you plant small flowering trees and shrubs, which are the walls, and you finish off with the fill-in plants—vines and ground-cover plants—which give your arrangement a finished look, like the roof and the paint job on your house.

Of course, you don't need to plant in this order—trees first, shrubs next, vines and cover plants last—but trees and many shrubs take at least several years of growth to make a showing. The sooner you plant them, the sooner they will be in flower and fruit to add to the beauty of your yard. Another big point in their favor—once your trees and shrubs are growing well, you won't need to give them the attention or maintenance you must give to your flowers and lawn.

**EASTERN PHOEBE**

| | |
|---|---|
| Number of Eggs Laid in Clutch: | 3-8 |
| Days to Hatch: | 14-17 |
| Days Young in Nest: | 15-16 |
| Number of Broods Each Year: | Usually 2, sometimes 3 |
| Lifespan: | 8 to 9 years |

*The eastern phoebe ranges throughout southern Canada, east to the Atlantic coast, west to the Rockies, and south to New Mexico. It nests in a cup of mud and mosses lined with grasses and hairs in the recess of a rock ledge, in caves, often on beams under bridges, on rafters in barns, porches, sheds, tops of shutters, or on a nesting shelf built for it.*

## HOW WE BEGAN

A number of years ago my wife and I were faced with the biggest gardening problem of our lives. We wanted to plant—in our Long Island backyard—trees, shrubs, and vines that we knew would especially attract birds. To do so, we would have to dig up and discard many shrubs that had already been planted there, which had little or no value to birds. We would also have to move others that we wanted to keep, to make room for the shrubs in our new planting plan. It was a big job, far bigger than planting a yard where nothing was growing, but we wanted robins, thrushes, catbirds, thrashers, song sparrows, and other birds that visited our feeding stations to remain in our garden and nest there.

I won't go into the details of what shrubs we already had in the yard, and what we moved, because it would only confuse you as it confused us in the beginning. We decided to close our eyes to what we had, and to draw a plan of what we wanted, just as if our yard didn't have a tree or a shrub in it. In this way we were starting fresh, like a youngster with a new notebook on his first day of school.

My wife wanted to plant immediately several flowering dogwood trees, which have showy white blossoms in spring, dark green leaves throughout the summer, and red leaves and red fruits in the fall. She couldn't have decided on a better small ornamental tree for a small yard, or a better one to attract birds. Almost a hundred different kinds of birds feed on the fruits of the various tree and shrub dogwoods that grow west to the prairies and beyond to the Pacific coast. A few dogwood trees in our yard would attract lots of birds when the fruits were ripe in the fall. But we also wanted, besides dogwoods, other trees and shrubs that would provide birds with fruits or seeds in early and late summer, and throughout the winter.

Fortunately I knew from government research which plants attract birds most to their fruits, not only in our suburban New York City backyard but in gardens across the country to Oregon, south to California, and east to the prairies, Texas, Florida, and Georgia. But first, before we planted anything, we had to make a plan, to put down on an outline drawing of our yard the exact *kinds* of trees, shrubs, and vines we wanted, how *many* the size of our yard would permit, and *where* they should be planted to grow best and to look most pleasing to us. Looking back on that first plan, I think we got as much fun out of it as we did in making our plantings.

**RULE OF THUMB**

The only way to ensure an orderly planting is to first plan it out on paper.

## *Starting Our Plan on Paper*

We discovered that there is only one way to make an orderly planting. To do so you must figure it out on paper, before you start to plant. First we measured our yard to find out how much room we had. We found it to be 60 feet wide, from neighbor to neighbor, and 140 feet deep, from the beginning of our front lawn to our back property-line fence.

After we had measured our yard, we next had to draw our property lines of a size on paper that would, proportionately, represent the actual size of our yard. To do this, we used a 12-inch ruler, and let each space between every one of the little marks on our ruler represent a foot (there are 16 of them to an inch on most rulers). Then we could lay out our 60 x 140 foot plot quite neatly within the boundaries of an ordinary sheet of typing paper.

When we had our outline of the yard drawn, we began to choose our trees, shrubs, and vines from the list of those that are most favored by songbirds in the Northeast. A map of our yard, followed by notes on what we planted, begins on page 192.

**ROSE-BREASTED GROSBEAK**

| | |
|---|---|
| Number of Eggs Laid in Clutch: | 3-5 |
| Days to Hatch: | 12-13 |
| Days Young in Nest: | 9-12 |
| Number of Broods Each Year: | 1 |
| Lifespan: | 4 to 9 years |

*The rose-breasted grosbeak ranges throughout the northeastern United States and into southern Canada. Its loose, cuplike nest of twigs is usually found 5 to 15 feet up near a crotch or fork of a small tree in woods, thickets, and in or near gardens.*

## WHAT FRUITS DO BIRDS LIKE?

For many years I had been interested in the food habits of songbirds, particularly the wild fruits that they like to eat. In the Adirondack Mountains of New York one winter day I saw a flock of big reddish pine grosbeaks eating the scarlet fruits of American mountain ash. At another time I remember fifty cedar waxwings that spent a cold stormy winter in the shelter of a grove of dark-green cedar trees in southern New Jersey. There every day I watched them eat the pale blue fruits of these trees. I have no doubt that the birds stayed in the grove not only because of the warm shelter of these evergreens but for the rich supply of cedar berries, which they lived upon all that winter.

In August and September in New York City's Central Park, I saw migrating robins, catbirds, cedar waxwings, rose-breasted grosbeaks, and thrushes strip wild cherry trees of their glossy black fruits. Later, in October, catbirds, brown thrashers, robins, and thrushes ate the scarlet berries of Amur honeysuckle, a tall ornamental shrub introduced into this country many years ago from Manchuria and Korea, which the parks department has planted there.

### How Scientists Discovered What Songbirds Eat

In government laboratories in Washington, D.C., beginning in 1883, men began examining the contents of birds' stomachs. Their researches were mainly to discover what services the birds were doing for mankind by eating certain insects, rats, mice, and other creatures that feed upon farm crops. Along with the insect foods of birds, they discovered the kinds of fruits and seeds that birds eat. Professor F. E. L. Beal, one of those early investigators, personally examined the contents of more than 37,000 birds' stomachs. W. L. McAtee, a younger bird food-habits expert, not only took over much of Professor Beal's work in the *Biological Survey*, but wrote many government bulletins about the food habits of birds, and how to attract them.

The results of sixty-five years of U.S. government research in wild-animal food habits have been compiled by several government scientists in a book, *American Wildlife and Plants*, published by the McGraw-Hill Book Company, New York. This book is so complete that it is likely that people will refer to it for years to come, whenever they want to learn what plant and insect foods are eaten most by songbirds and by other kinds of wild animals throughout the United States. My lists of trees, shrubs, and vines to attract birds, in this chapter, are based on the regional food habits of songbirds given in American Wildlife and Plants.

I had other records of native and foreign plants—elderberries, hollies, dogwoods, and viburnums—whose fruits I had seen songbirds eating. But if I were to attract the most birds, I could not choose the plants for my own yard based upon those few observations. I needed scientific information—thousands of records of bird food habits—such as the Fish and Wildlife Service, formerly the U.S. Biological Survey, had been gathering in many parts of the United States for more than fifty years. These would truthfully show the seasonal foods of birds—not only the berries and fleshy summer fruits, but the dry, hard seeds of birches and pines and the acorns of oaks that songbirds eat. Knowing what they favored through each season, I would be sure of planting the trees, shrubs, and vines that would offer them a food supply the year around.

## PLANTS BIRDS NEED YEAR ROUND

**RULE OF THUMB**

Remember to pick out trees, shrubs, and vines that will provide birds with year-round food, shelter from the elements, and summer nesting places.

We looked on our choice of plants for the birds in our yard somewhat in the way we would provide for our own needs. We knew that the greatest use birds get out of trees, shrubs, and vines are for: (1) food, (2) shelter from cold and snow and from summer storms, and (3) nesting places. If we could provide the three basic needs for songbirds—year-round food, winter shelter, and summer nesting places—we had no doubt that they would come to our yard. We were also sure that when our planting had a good start, we would have birds in our garden every day of the year.

We had only one other consideration before we selected our list of plants. We would also try to choose those that would provide colorful fruits, flowers, leaves, and interesting outlines against the sky, even in winter when the leaves and fruits of some of them are gone. This wasn't difficult because most trees and shrubs have some special attraction at some particular time of the year.

## SOURCES OF HELP

**TRY THIS**

A landscape architect who is familiar with the plants preferred by birds of your region can also draw up a planting plan for you.

If you are too busy to make your own planting plan, I suggest that you consult a landscape architect, who, for a fee, will draw a planting plan for you. Be sure to make it clear that you want him to choose plants for your yard from the list of those preferred as food by the birds of your region. These lists are given, by regions, later in this chapter. Most of these plants, whether they are native or introduced from other countries, are excellent ornamentals, besides their food, cover, and nesting values to songbirds. Your landscape architect should be able to draw a beautiful planting plan for your garden. He should also be most capable of recommending the species or kinds of trees, shrubs, and vines on your regional plant lists in the Appendix that will grow best in your local area.

Or you might consult the county agricultural agent (there is such an office in the county seat of all U.S. rural counties) about what you would like to do. If he does not feel capable of advising you, he may be able to suggest other sources of help. Your state college of agriculture will have free bulletins or leaflets that will describe some of the trees, shrubs, and vines that grow in your state. Soil conservationists and federal and state foresters may be willing to tell you what species of trees and shrubs on your regional list will grow best in your community. Perhaps the owner of your local nursery may be willing to help you to plan and plant your garden to attract birds.

TRY THIS

Your county agricultural agent might also help advise you of which plants will grow best in your community.

## *Deciding When to Plant*

Before we started our new songbird planting plan, we spent a lot of time, from spring until fall, on the terrace just in back of our house. It was always pleasant sitting there. On clear days, we had sunshine until noon, and when the sun got over into the west, our house shaded us from the afternoon heat. Sitting in comfortable chairs on the terrace, or lying in the swing, we could look out over our flower beds and lawn clear to our back property-line fence. It was a pretty sight, and if we kept this view open, we could even improve upon it with our bird plantings.

To keep our view from being obstructed, we decided not to plant any trees or shrubs between the terrace and our back lawn, but to concentrate them along the sides of our property and back fence. I was particularly eager to plant the back line anyway. Just over the line, we faced the backs of two cement-block garages in a neighbor's yard. They weren't pretty to look at, and our back line planting would hide them from our sight.

But first, how far in upon the lawn from our property lines did

RULE OF THUMB

You should maintain a depth of at least 6 feet of shrubbery from fence line to lawn to give songbirds enough privacy and room to nest.

TRY THIS

To get a better idea of how a planting line will look in your yard, mark it off on the ground with a length of hose or clothesline.

we want to plant? To give songbirds room to nest and nesting privacy, we knew that they needed a depth (or width) of at least 6 feet of shrubbery from fence line to lawn, and a greater width wherever possible. We decided then that the narrowest part of our shrub border, at any place along our property line, would be at least 6 feet wide.

Curving lines in a planting look good, and we had to decide upon an inner line in the yard where shrubs and lawn would meet. Like most people, we wanted to see this line *on the ground* before we decided where it was to be permanently. We took several lengths of garden hose and laid them out in the curve that you see outlined by our flagstone walk on the map (page 192). The line marked by the hose (a clothesline would have done just as well) looked graceful, and it allowed a planting depth, or width, from lawn to fences that would enable us to plant a variety of trees and shrubs that birds like. We knew that the more of the different kinds of their preferred plants we could use, the more attractive our yard would be to a greater variety of birds.

To mark permanently our line where the future shrub planting would meet the lawn, we bought two dozen flat pieces of flagstone. These we set at ground level just below the grass, about every six to eight feet apart along the line marked by the hose. Eventually, we would buy enough flagstones to set them about every two feet apart along this line. For the present, these would mark the future boundary between our shrubbery and lawn. They were the beginning of our flagstone walk that someday would circle completely about our backyard, and serve to keep us and our guests from damaging the grass by walking on it too frequently.

# *Putting the Plants on Your Map*

The "needle" evergreens—pines, cedars, spruces, yews, junipers, hemlocks, etc.—are the backbone of any planting for songbirds in the northern states and in many parts of the South where winter cold spells may occasionally be severe. Broad-leafed evergreens such as rhododendrons, though effective cover, do not have as many attractions for birds. The evergreens help songbirds by providing them with protective shelter throughout the year. Of the two hundred trees and shrubs we planted in our yard, about 9 percent, or eighteen of these, were "needle" evergreens. After you have made your preliminary planting plan on paper, add up the total number of your plants and see how many of them are pines, spruces, yews, etc. If they are from 8 to 12 percent of your total number of plants, you may feel assured that you have planned adequate shelter for the birds in your yard.

RULE OF THUMB

To ensure that your birds have enough shelter during winter months, 8 to 12 percent of your total number of plants should be evergreens.

Most evergreens, except the dwarf varieties, grow into tall trees in time. Assign them places in the *back* of your planting, in *corners* of your yard, and in places along the *sides* of your property. There they will make an effective green wall, or a contrasting background for your flowering and fruiting trees and shrubs.

We put some of our cedars along our north and east property lines (see A on the map of our planting plan, page 192). They were directly back of our white-flowering dogwoods and white-blossoming serviceberry (at 5 and 7). We used pines (at B) as a green background for the white trunk of a birch tree (at 9). By planting food-producing trees—dogwoods, serviceberry, and birch—in front of and close to the evergreens to get an ornamental effect, we provided food for birds near the safety of cover. The evergreens, with their densely needled branches, were a refuge where songbirds could escape from the small hawks or shrikes that occasionally tried to catch them while they were engrossed in feeding.

To make the cover almost immediately useful to birds, without

**TRY THIS**

Space your cedars and white pines close enough together to provide thick cover for the first two years.

the necessity of having to wait five to ten years for the trees to spread out, we spaced our cedars (and white pines, too) close enough to each other to provide thick cover the first year or two after we had planted them. We did this while quite aware that perhaps in ten or fifteen years we might have to transplant some of the evergreens. By that time they would have broadened, with their branches intermingled and possibly their growth slowed a bit. By planting them close together, we nevertheless achieved our purpose of providing "quick" cover for birds during those first years immediately after planting.

## How Far Apart Should You Space Trees and Shrubs?

**RULE OF THUMB**

Plant your taller trees a distance apart approximately equal to one-third of their ultimate height. Smaller trees and tall shrubs should be planted a distance equal to between one-half and two-thirds of their maximum height.

We planted our tree evergreens and most of our small flowering trees much closer together than a landscape gardener or a nurseryman would recommend. Our dogwoods, serviceberry, fire cherry, and mulberry we spaced about 8 to 10 feet apart. This was approximately one-third the ultimate height of about 30 feet to which most of these trees grow. The majority of small trees and tall shrubs are usually planted a distance apart about equal to one-half to two-thirds their maximum height. For example, flowering dogwood trees—which may eventually grow 30 feet tall or more—should normally be spaced at least 15 to 20 feet apart, instead of 8 to 10 feet apart, as we spaced ours. You may prefer to plant yours wider apart than we did, and you will probably have a better-looking planting than ours, after fifteen or twenty years have passed.

**TRY THIS**

Larger dogwoods that grow to a height of between 12 and 15 feet will help fill in the wider spaces between plantings.

If you don't like the idea of having openings that will show up in your wider-spaced planting, you might, before you plant, consider buying larger dogwoods—say 12 to 15 feet tall—instead of the 8- to 10-foot sizes that we chose. These will have wider-spreading branches because of their larger size, which will help to fill in the spaces between them. If your budget won't allow you to buy the

### Virtues of the Homely Pokeberry

Since the first edition of *Songbirds in Your Garden* was published, many of my friends and other readers of this book have written to ask why I did not include pokeberry, or pokeweed, as a recommended bird-food plant for the garden. Its purple-black berries on red stems are eaten by fifty-two kinds of birds.

In listing ornamental plants that attract birds, I did not include pokeberry because of its homeliness. Variously called pokeberry, pokeweed, poke, scoke, garget, redweed, and pigeonberry, the botanical name of this plant is Phytolacca americana. Many people consider it an ungainly weed in the garden; but, if allowed to grow in an inconspicuous place, such as a corner where it is not easily seen, it will add much to your garden's attractiveness to birds. It grows naturally in sunny, moist places, along the edges of woods, fields, roadsides, streams, and fence corners, from Maine to Florida, west to Texas and north through the Plains and Prairie states of Oklahoma, Missouri, Kansas, Iowa, and Nebraska to parts of Minnesota and Wisconsin.

I don't know of any nursery that sells the plant because it is considered a weed, but it usually is planted by the birds themselves. After eating its berries they drop its black seeds wherever they happen to travel. Mourning doves are especially fond of pokeberries; they are also favored by bluebirds, catbirds, woodpeckers, robins, mockingbirds, thrushes, vireos, cardinals, cedar waxwings, and many other garden birds.

For the ten years that we lived on Long Island, a pokeberry grew up freshly each spring from its big root in a sunny corner in the back of our garden (it is a perennial herb that dies back to the ground each fall). One year, in its prime, our eleven-foot leafy green plant grew a crop of berries that I estimated at ten thousand. The crop ripened in August. Within a week, our garden birds had eaten every one of the berries from their favorite bush.

**MOURNING DOVE**

| | |
|---|---|
| Number of Eggs Laid in Clutch: | 2-4 |
| Days to Hatch: | 14-15 |
| Days Young in Nest: | 14-15 |
| Number of Broods Each Year: | 2 to 5 |
| Lifespan: | 5 to 10 years |

*Mourning doves reside in every state in the United States. Their nests are found on platform sticks in tree crotches and on branches usually 5 to 25 feet up, also in vines, on stumps, roof gutters, and in garden arbors.*

larger and more expensive sizes and you still want the wide spacing between your trees, you might consider buying some inexpensive low shrubs—perhaps coralberry or snowberry, for example—and plant these between the trees. These will provide fruits that birds eat. Eventually your trees will "shade out" these shrubs, but in the meantime, they will serve their purpose of filling up the open spaces and providing food.

If you don't want to use shrubs to fill in, you can plant a low-growing, permanent ground cover of partridgeberry, bearberry, or wintergreen. Each of these has red fruits eaten by birds. If they are difficult to find in nurseries, you can buy English ivy, *Hedera helix*, or its varieties, Japanese spurge, *Pachysandra terminalis*, or some other trailing or low-growing plant. (Don't use Japanese honeysuckle, *Lonicera japonica*, or its varieties. This is a vigorous, climbing vine that will overrun your small trees if it is planted under them.)

All of these small broad-leaved evergreens will keep out the weeds by covering the bare ground, and are attractive to look at all year. Eventually, we chose pachysandra for a ground cover under our flowering trees and shrubs. Many nurseries sold it, and it gave a neat, luxuriant finish to our plantings that was especially needed on the inside border, next to the lawn, which was in view of anyone walking about in our yard.

TRY THIS

A low-growing, permanent ground cover of partridgeberry, bearberry, or wintergreen is a good substitute for shrubs to fill in gaps in your planting.

When spacing your shrubs on your planting plan, you can put them closer together, in proportion to their height, than trees. If you are planting a thicket of shrubs to attract birds, as we did with gray dogwood (at 11 on the map) and with arrowwood (at 14), you can set your shrubs about 2 feet apart. When you plant vines at the bases of walls, arbors, and trellises, you might set them close together, as with shrubs, if you want quick results in getting a thick cover.

RULE OF THUMB

Space shrubs closer together, in proportion to their height, than you would trees.

## LARGE TREES IN YOUR YARD

You may have wondered whether or not your yard is large enough to accommodate a large tree. No matter what tree you buy, or how long it may take it to reach its full height—some oaks and other slow-growing trees may take at least a hundred to two hundred years—you should consider its ultimate height in proportion to the size of your yard and house.

Our yard was 60 feet wide and 140 feet deep—probably larger than many suburban lots of today. Our house was two floors and an attic high. The white oak tree, near the center of our backyard, about 70 feet tall, was neither too big for the property, nor too much out of proportion to the house. Had our house been a one-story-high ranch design, we might have considered our oak to be too tall. The oak grew there long before the house was built. Had we planted it, we would have put it in a back corner of the yard, or somewhere along the north property line, where it wouldn't have cast a large shadow in summer over the center of the yard, which it did while we lived on Long Island. A tree that doesn't get too tall is far better for a low house than one that, someday, will dwarf it.

RULE OF THUMB

**Consider a tree's ultimate height in proportion to the size of your yard and house before you buy.**

If you want a fine shade tree for a small yard—a tree that will endure city smoke, dry, poor soils, and attract birds—try the sugar hackberry, *Celtis laevigata* (pronounced SELL-tiss leave-ih-GAY-tah). It grows, after many years, to 75 or 80 feet tall, and has a dark red or purple fruit that looks like a berry. These are ripe in September and October and sometimes stay on the tree through the winter. About forty-seven different kinds of birds eat these fruits, including cardinals, cedar waxwings, flickers, mockingbirds, robins, and olive-backed thrushes. The sugar hackberry grows from southern New England west through Pennsylvania and Ohio to Indiana, south to Texas, and east to Florida. It looks like an elm and is a good substitute for it. Another hackberry, *Celtis occidentalis* (pronounced SELL-tiss ock-sih-den-TAY-liss), is hardy north into Canada and is much planted for windbreaks in the Plains and Prairie region of the United States. Unfortunately it is susceptible to the "witch's broom disease," which doesn't seriously harm it, but forms rather unsightly clusters of twigs on the tree. If you don't mind this defect, this hackberry is hardier than the sugar hackberry and is slightly more attractive to birds.

TRY THIS

**The sugar hackberry is a good, tenacious shade tree that will attract birds to a small yard.**

Another, smaller, tree that also endures city smoke and poor, dry soils, and attracts birds, is the Amur cork tree, *Phellodendron amurense* (pronounced Fel-low-DEN-dron ah-moor-EN-see). In

New York City's Central Park, I have seen flocks of migrating robins, catbirds, and thrushes sit in one of these trees day after day through October and November, eating the black, grapelike fruits as long as they were available. The tree grows to about 45 feet tall, is wide and spreading, with massive branches and rough bark. The sexes are separate so you must plant both a male and a female tree if you want fruits for birds. The Amur cork tree is not a large tree, and a property the size of ours would have accommodated two of them, if we had no other large trees in our yard.

If we were using this tree in our garden, we would plant the male tree on our terrace, near the house, and the female tree, which bears the black, juicy fruits, in the back of the yard. There, in late fall when the fruits are ripe, they would stain no sidewalks when they fell to the ground. This hardy tree will grow from eastern Canada west to Montana in the United States, and southward.

If you have a small yard, you should also consider planting a blue beech, *Carpinus caroliniana* (pronounced Car-PIE-nus car-ol-inna-AY-nah). This small native tree grows to about 35 feet tall and has leaves resembling those of an elm. Its hard nutlets, hanging from the tree in small leaflike clusters, attract cardinals, myrtle warblers, and other birds, which like the seeds. In Central Park, where these trees grow commonly, I have watched gray squirrels in autumn feed on the little clusters of seeds day after day until they have eaten them all from the trees. This is one of the hardiest trees in North America and grows from far up in the colder parts of Canada south to Texas and Florida.

In the South, particularly in Florida, where it is known as Christmas-berry tree, people plant the handsome, 40-foot-tall Brazil pepper tree on their lawns and along the city and town streets. In California another species, the California pepper tree, is much planted as a highway or street tree (however, it must not be planted near orange groves because it attracts the black scale—an insect that also attacks citrus trees). In California, bluebirds, mockingbirds, phainopeplas, robins, and other birds eat the abundant red berries that grow in clusters on this tree in the fall.

The Pacific madrone, or madroña (pronounced ma-DRONE-yah), *Arbutus menziesi*, and the strawberry tree, *Arbutus unedo*, are two handsome broad-leaved evergreens with brilliant red to orange berries that birds eat. Both of these are interesting and attractive shade trees for small yards in the Pacific region.

**RULE OF THUMB**

Plant your shade trees near the back of your yard or garden or along one of its sides so that they won't shade the center of your lawn.

When you decide on the kind of shade tree or trees that you want in your garden—whether a tall oak, elm, or maple or a dogwood, cherry, or other small tree—mark it on your map of your yard near the back or along one of its sides. After the tree is planted, it will look better there and won't shade the center of your yard, which is usually more attractive if kept an open lawn.

After you have chosen your species of trees, shrubs, and vines, visit your nearest nursery to see if it has in stock the plants you want. If you go there in spring or fall, the busiest time, don't expect to get the personal attention that you might be able to get at other times of the year. We liked to visit our local nursery in summer when leaves, flowers, or fruit are on plants and we can "browse around," as some people do in a bookshop, without feeling that we are in the way. Our nurseryman knew us so well that he tried to let us know we were always welcome, even though we might only have been "window shopping."

**TRY THIS**

Summer is the least crowded time to browse at your local nursery.

It is best to get well acquainted with your plants before you buy them—to know how they look at different times of the year, their ultimate height, rate of growth, and time of flowering or of fruiting. We don't know of any way to get better acquainted with plants than to visit a nursery, unless you have friends who have planted the kinds of trees and shrubs you are interested in, or know someone who owns, or superintends, a large estate. If you live near an arboretum, or botanical garden, one of those wonderful places where trees, shrubs, and other plants are grown especially for people to see and for scientists to study, you are fortunate. Go there as often as you possibly can, and you will be pleased at how soon and how much you will learn about plants.

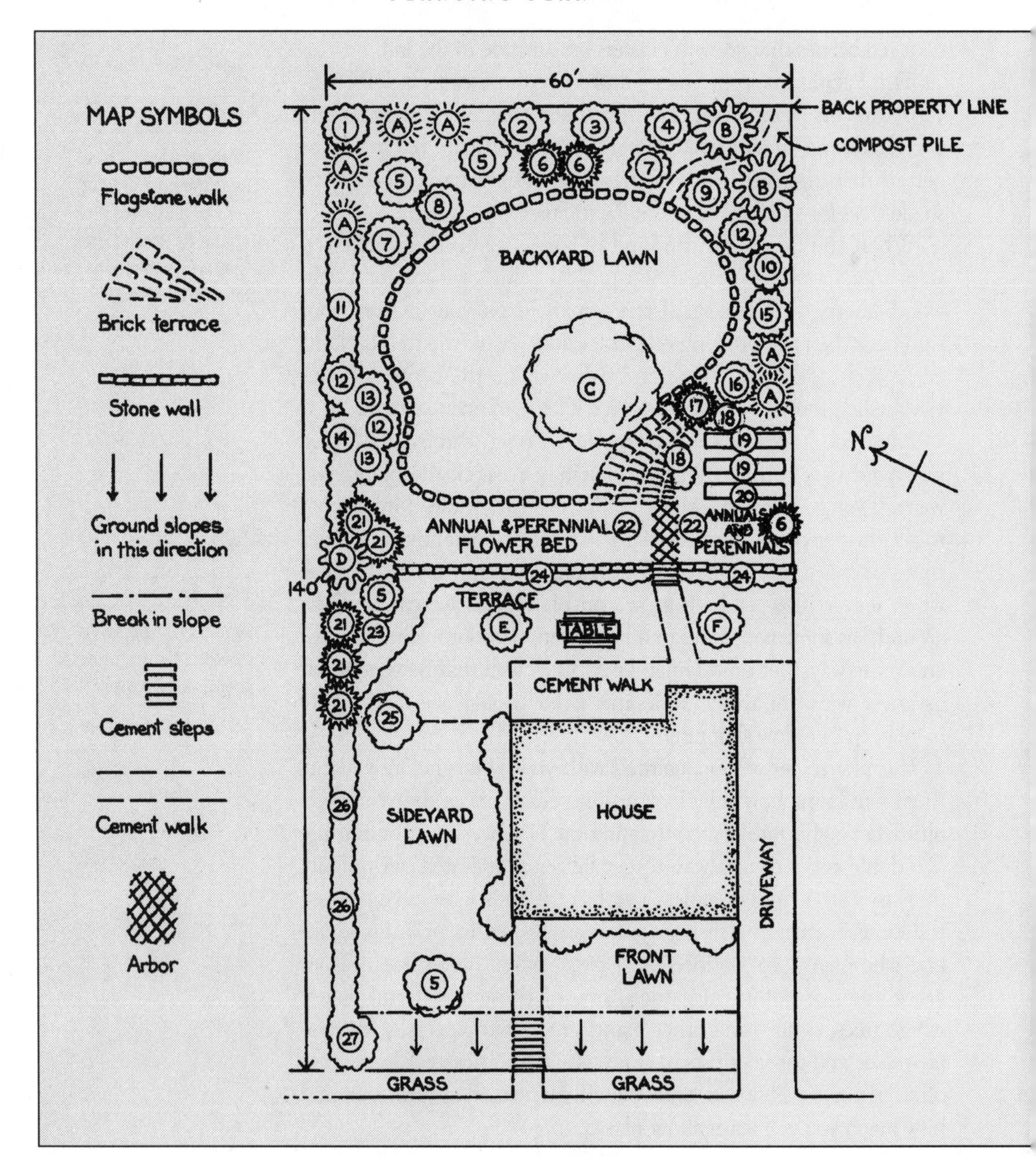
MAP SYMBOLS
Flagstone walk
Brick terrace
Stone wall
Ground slopes in this direction
Break in slope
Cement steps
Cement walk
Arbor
60′
140′
BACK PROPERTY LINE
COMPOST PILE
BACKYARD LAWN
ANNUAL & PERENNIAL FLOWER BED
ANNUALS AND PERENNIALS
TERRACE
TABLE
CEMENT WALK
SIDEYARD LAWN
HOUSE
FRONT LAWN
DRIVEWAY
GRASS
GRASS
N

## Key to Our Long Island (New York) Planting Plan

### LARGE TREES

A Red cedar (*Juniperus virginiana*)

B Eastern white pine (*Pinus strobus*)

C White oak (*Quercus alba*)

D Red pine (*Pinus resinosa*)

E American, or white, elm (*Ulmus americana*)

F Box elder, or ash-leaved maple (*Acer negundo*)

### SMALL TREES, SHRUBS, AND VINES

1 Russian mulberry (*Morus alba tatarica*)

2 Fire, or pin, cherry (*Prunus pennsylvanica*)

3 Washington hawthorn (*Crataegus phaenopyrum*)

4 American elder (*Sambucus canadensis*)

5 Flowering dogwood—white (*Cornus florida*)
Flowering dogwood—pink (*Cornus florida rubra*)

6 Thayer's yew (*Taxus cuspidata thayerae*)

7 Allegany serviceberry (*Amelanchier laevis*)

8 Maries doublefile viburnum (*Viburnum tomentosum mariesi*)

9 Gray, or poverty, birch (*Betula populifolia*)

10 Siebold's viburnum (*Viburnum sieboldi*)

11 Gray dogwood (*Cornus racemosa*)

12 Orange-fruited viburnum (*Viburnum setigerum aurantiacum*)

13 Linden viburnum (*Viburnum dilatatum*)

14 Arrowwood (*Viburnum dentatum*)

15 Amur honeysuckle (*Lonicera maacki*)

16 Japanese dogwood (*Cornus kousa*)

17 Dwarf Japanese yew (*Taxus cuspidata densa*)

18 Japanese barberry (*Berberis thunbergi*)

19 Blackberries and raspberries (*Rubus*)

20 Highbush blueberries (*Vaccinium*)

21 Japanese yew (*Taxus cuspidata*)

22 Trumpet creeper (*Campsis radicans*)

23 European cranberry-bush (*Viburnum opulus*)

24 Virginia creeper (*Parthenocissus quinquefolia*)

25 Sargent crabapple (*Malus sargenti*)

26 Black haw (*Viburnum prunifolium*)

27 Small-leaved cotoneaster (*Cotoneaster microphylla*)

## THE TREES AND PLANTS IN OUR GARDEN

Depending on where in the country you live, and how large your yard is, your plantings will be quite different from ours, but it may be helpful to see some of the factors we considered in planting our yard. Near the end of the chapter you will find a list of trees and shrubs, listed according to region, that provide food and shelter for birds. You can use that list to help you decide what to plant in your own garden and yard.

### LARGE TREES

**A. Red cedar** (a large tree, which grows to 90 feet when mature) is one of the best of all evergreens for songbirds because it provides food, shelter, and nesting cover. It is also a fine ornamental, but is susceptible to rusts, which it can transmit to apples, hawthorns, mountain ashes, juneberries, and quinces; these can be controlled on red cedar by spraying in early in May with a 1-percent solution of Fermate or Elgetol. It is best not to plant red cedars near apple orchards.

**B. Eastern white pine** (a large tree, which grows over 100 feet when mature) is one of the most beautiful of all pines, and is particularly useful as a background tree for flowering trees and shrubs. A favored nesting tree of robins, blue jays, and mourning doves, it is susceptible to blister rust, which is transmitted to white pine from currants and gooseberries. Blister rust can be controlled on individual trees, but they will be reinfected if currants and gooseberries are growing nearby.

**C. White oak** (a large tree, about 90 feet when mature) is slow-growing, but one of the longest-lived, sturdiest, and most picturesque of all oaks. It bears a crop of acorns each year, although the size of the crop may vary considerably. The black oak group, which includes the rapid-growing northern red oak, scarlet oak, pin oak, and others, requires two years to mature its acorns, instead of one year.

**D. Red pine** (a large tree, about 75 feet when mature) is handsome, rapid growing, and generally resistant to diseases. This hardy pine is adaptable to most of the Northeast region. There is a dwarf, rounded form, called variety *globosa,* that was discovered in New Hampshire about the year 1910.

**E. American, or white, elm** (a large treee, about 120 feet when mature) is one of the handsomest of all American trees. Orioles nest in its end branches, finches and grosbeaks eat its buds and seeds. Unfortunately, it is susceptible to both the deadly Dutch elm disease and phloem necrosis, for which there is, at present, no cure.

**F. Box elder, or ash-leaved maple** (a large tree, about 60 feet when mature) is not a good ornamental, but its winged seeds are so important to wintering evening grosbeaks that one should be included in every songbird planting in the Northeast region.

### SMALL TREES, SHRUBS, AND VINES

**1. Russian mulberry** (a small tree, about 25 feet when mature) is not an attractive ornamental, but one of the very best trees to attract birds in early summer. Birds are so fond of mulberries that they will pass up cultivated cherries, strawberries, and other fruits in favor of them. Every yard should have a mulberry tree, but you may need both a male and female tree to be sure that the female tree will set fruit. Do not plant mulberry trees near your sidewalks or in places where you might sit under the tree. The fruits that fall will stain your clothing and discolor your walks.

**2. Fire, or pin, cherry** (a small tree, about 35 feet when mature) is a better tree for the small yard than its much larger relative, the wild black cherry, *Prunus serotina.* The fire, or pin, cherry is extremely hardy, and grows from Newfoundland south to North Carolina, and west to Colorado. At least twenty-three kinds of birds eat its small red cherries. It flowers in early May in New York State, and has beautiful red shining bark.

**3. Washington hawthorn** (a small tree, about 30 feet when mature) has the thorny branches and dense foliage that make hawthorns preferred nesting trees for many birds. Fruit persists on the branches of Washington hawthorn through the winter and makes good emergency bird food. This is one of the best ornamentals of all the hawthorns and is the one most resistant to the cedar-hawthorn rust.

**4. American elder** (a tall shrub, about 12 feet when mature) is not a particularly attractive ornamental, and rather coarse leaved and vigorous growing, but birds are very fond of its purple or blue-black berries. The white flower clusters are attractive and bloom when most other shrubs have finished flowering. Try to buy the variety *maxima,* which has much larger flower clusters. Some of these may be 12 inches to 15 inches in diameter.

**5. Flowering dogwood** (a small tree, 15 to 35 feet when mature) is the finest ornamental tree of all natives in the northern United States. It grows slowly, is long-lived, and will live either in sun or shade, dry or well-drained soils. It is usually best to buy good sturdy nursery-grown specimens, rather than to transplant it from the wild. The pagoda dogwood, *Cornus alternifolia,* is a good substitute for it and grows farther north.

**6. Thayer's yew** (an evergreen shrub, about 8 feet when mature) provides roosting, nesting, and wintering cover for birds. Attractive red fruit appears on the *female* plants. For every five female plants, buy one male, or staminate, plant, to be sure that female, or pistillate, plants bear fruit. Thayer's yew, and variety *densa,* are low and wide-growing, which makes them excellent for small yards.

**7. Allegany serviceberry** (a small tree, about 35 feet when mature) is better for the small yard than downy serviceberry (*Amelanchier canadensis*), which grows much taller and is commonly sold by nurseries. If you can find it in a nursery, try to get *Amelanchier grandiflora,* which is a cross between *laevis* and *canadensis.* It is a smaller tree (up to 25 feet) than either of its parents, and has larger flowers. Birds like the fruits of them all.

**8. Maries doublefile viburnum** (a medium shrub, about 9 feet when mature) produces beautiful white flowers that bloom shortly after flowering dogwood has finished. Red berries that birds eat appear in summer. It may grow as wide as tall; the 'Maries' variety has larger flower clusters than *V. tomentosum,* the species itself.

**9. Gray, or poverty, birch** (a small tree, about 30 feet when mature) is a particularly beautiful tree. Besides the seeds it offers chickadees, redpolls, and pine siskins, it has a tremendous attraction for the small, brightly colored warblers. These birds come to the tree to feed upon birch plant lice, which occasionally infest birches. When these insects are numerous on these trees, they will attract all of the local warblers, and the migrating ones, too, in spring and fall.

**10. Siebold's viburnum** (a small tree, about 30 feet when mature) is one of the most attractive of all viburnums and makes a splendid specimen plant in the garden. Beautiful in flower and fruit, birds like its red and black berries, borne on red fruit stalks in late summer and early fall. It is hardy in most northeastern states.

**11. Gray dogwood** (a tall shrub, about 15 feet when mature) is a dense, native shrub that endures city smoke and withstands shearing or clipping. It can be cut back to 6 or 8 feet high, and it makes a good thicket for birds to nest or hide in. Its white berries on red stalks are ornamental, and at least twenty-two kinds of birds eat them. It will grow in dry, well-drained, or moist soils, and in the sun, or in shade.

**12. Orange-fruited viburnum** (a tall shrub, about 12 feet when mature) is the only viburnum with beautiful orange-colored fruit. We have watched migrating hermit thrushes in our backyard pull the orange berries from our bushes to eat them. The berries remind one of mountain ash berries.

**13. Linden viburnum** (a medium shrub, about 9 feet when mature) is one of the best of the ornamental viburnums for its abundant bright red fruit. Migrating thrushes eat the berries from our bushes in the fall.

**14. Arrowwood** (a tall shrub, about 15 feet when mature) is a fine ornamental for city or suburban yards; it is hardy, grows in almost any soil, and endures city smoke. Its dense green foliage turns glossy red in fall. The blue berries are eaten by several kinds of songbirds, and thickets of this shrub make good nesting cover.

**15. Amur honeysuckle** (a tall shrub, about 15 feet when mature) is very hardy, possibly as adaptable to poor, dry soils and cold weather as Tatarian honeysuckle, *Lonicera tatarica.* A nursery as far north as Manitoba, Canada, raises Amur honeysuckle. We have seen migrating thrushes, robins, catbirds, and brown thrashers eating the red berries of this shrub in late October, when few other shrubs in our neighborhood have juicy fruits on them. We prefer the Amur honeysuckle to Tatarian honeysuckle, because its fruits are ripe at a time of fruit scarcity for small birds in our area.

**16. Japanese dogwood** (a small tree, about 20 feet when mature) attracts birds to eat its red, raspberrylike fruit. Its beautiful white flower bracts appear after flowering dogwood has finished blooming. It is not supposed to be hardy, or able to stand the winters north of southern New England, Pennsylvania, Ohio, southern Indiana, southern Illinois, and Missouri.

**17. Dwarf Japanese yew** (an evergreen shrub, about 4 feet when mature) is a fine dark-green ornamental that is hardy in most of the northeastern states and will grow in many different kinds of soils. They are raised in most American nurseries and are hardier than the English yew, *Taxus baccata,* and its varieties.

**18. Japanese barberry** (a medium shrub, about 7 feet when mature) is one of the most widely used hedge plants, and noted for its ability to grow in dry soils. Its dense, thorny growth, bright red fruits, which hang on all winter, and scarlet leaves in autumn, make it desirable in most gardens. It is a favorite nesting bush for song sparrows, at least in the Northeast region. At least seven kinds of birds eat its red fruits.

**19. Blackberries and raspberries** (both medium shrubs, about 9 feet when mature) are available at some nurseries that specialize in growing these and blueberry plants. Although some varieties grow over a great area, it is best to write to your state college of agriculture or your local county extension agent, *before you buy your plants,* and ask what varieties they recommend for your locality. Also ask them for a bulletin that will tell you how to plant and care for them.

**20. Highbush blueberries** (a tall shrub, about 12 feet when mature) are available at nurseries. We like varieties 'Weymouth' and 'June' for early fruit; 'Ivanhoe' and 'Stanley' for midseason; 'Atlantic,' 'Jersey,' and 'Coville' for late berries. Consult your state agricultural college or county agent before you buy your plants.

**21. Japanese yew** (an evergreen tree-shrub, 15 to 20 feet when mature) is a tall, upright yew variety called *capitata* by some nurserymen. It provides dark-green ornamental background for other plants, and good winter cover for birds. Be sure to plant both male and female plants if you want the attractive pink-red fruit.

**22. Trumpet creeper** (a vigorous vine, about 20 to 30 feet when mature) is an excellent hummingbird food plant that has scarlet, and orange, trumpet-shaped flowers in midsummer. A hybrid, *Campsis tagliabuana,* variety 'Madame Galen,' has larger, showier flowers and is said to be equally attractive to hummingbirds.

**23. European cranberry-bush** (a tall shrub, about 12 feet when mature), offers red fruit that persists sometimes until spring. The fruit is eaten by pine grosbeaks and other wintering birds. This plant seems to be hardy across the northern United States to the Mountain Desert region and beyond to the Pacific region.

**24. Virginia creeper** (a vine of between 10 and 20 feet when mature) features beautiful ornamental leaves in fives that turn scarlet in fall; its blue fruit is very attractive to many kinds of birds. This vine is healthy, but never grows so vigorously that it will be running wild in your garden as some vines might do.

**25. Sargent crabapple** (a small tree, about 8 feet when mature) is the smallest of the crabapples; no larger than a shrub, it gets as wide as tall. Resistant to the cedar-apple rust—a fungus disease that attacks red cedars and apple trees—it is beautiful in flower and in fruit.

**26. Black haw** (a small tree, about 15 feet when mature) makes a luxuriant hedge and single plants make good specimens. It is good for "screen planting" to give you privacy or to blot out an ugly view, and it provides fruit and nesting places for songbirds.

**27. Small-leaved cotoneaster** (a low shrub, about 3 feet when mature) is a handsome, shiny-leaved, low plant that is particularly good for embankments or in rock gardens. It provides red fruit throughout the winter that is available to birds. In California, people see birds frequently eating the red berries of the cotoneasters. In England, in the winter of 1949-50, bird scientists found Bohemian waxwings eating immense quantities of its fruit. Some of these birds ate far more than their own weight of the berries each day

## WHEN OUR TREES, SHRUBS AND VINES WERE IN FLOWER, AND THE AUTUMN COLORS OF THEIR LEAVES

The list does not include the red cedar, Russian mulberry, yews, and pines because their flowers are inconspicuous. The blooming period given is for southern New York and southern New England generally. Some of the plants on our list, when grown farther north, would bloom weeks later. Those grown farther south would bloom weeks earlier.

### APRIL

(early) American elm — not showy, leaves usually yellow-brown
(mid-) Box elder — not showy, leaves usually dull green to brown
(mid-) Gray birch — not showy, leaves bright yellow
(late) Serviceberry — showy white flowers, leaves yellow to red

### MAY

(early) Fire cherry — small white flowers, leaves red
(mid-) Japanese barberry — not showy, leaves scarlet
(mid-) Flowering dogwood — showy white flower bracts, leaves scarlet
(mid-) Sargent crabapple — showy pure white fragrant flowers, leaves green
(mid-) White oak — not showy, leaves purple red, violet, and sometimes rusty green-gold
(late) Black haw — white flowers in flat clusters, leaves shining red
(late) Seibold viburnum — cream-white flowers in flat clusters, leaves red
(late) Maries doublefile viburnum — cream-white flowers in flat clusters, leaves dull red
(late) Amur honeysuckle — fragrant white to yellowish flowers, leaves green
(late) Highbush blueberry — white or pinkish flowers, leaves scarlet

### JUNE

(early) Japanese dogwood — showy white to pink flower bracts, leaves scarlet
(early) Arrowwood — cream-white flowers in flat clusters, leaves glossy red
(early) Linden viburnum — showy cream-white flowers, leaves green to russet red
(early) European cranberry-bush — white flowers in flat clusters, leaves red
(mid-) Washington hawthorn — showy white flowers in many-flowered clusters, leaves scarlet to orange
(mid-) Small-leaved cotoneaster — not showy, small white flowers, leaves evergreen, shining
(mid-) Gray dogwood — cream-white flowers in flat clusters, leaves purple
(late) Common elderberry — showy white flower clusters a foot in diameter, leaves green to dull brown

### JULY

(early) Orange-fruited viburnum — white flowers in flat clusters, leaves green to russet red
(mid-) Trumpet creeper — handsome orange-red trumpet-shaped "hummingbird" flowers, leaves purplish brown

## THE MONTHS WHEN OUR PLANTS OFFERED FOOD TO BIRDS

Listed below are the times of the year during which our plants ordinarily held their fruits in our Long Island, New York, area of the Northeast region. During some years, our plants produced large crops, in other years, small ones.

In our garden, the special value of the red cedar, linden viburnum, European cranberry-bush, Japanese barberry, Washington hawthorn, Sargent crabapple, small-leaved cotoneaster, and box elder was their ability to hold their fruits and seeds on their branches for a long time. There, above the snow, they were available to birds during periods when they might have needed them the most. These fruits and seeds we called "emergency foods." Birds did not usually eat them until other foods were scarce, or covered by snow. Perhaps a certain amount of weathering of some of these fruits, like Japanese barberry, was necessary before they were palatable to birds. Plants of this kind may help many of the birds wintering in your garden to survive.

Unusually large flocks of birds would sometimes suddenly descend on these plants and strip them of their fruits long before the bearing periods of the plants had ended. Some individual plants dropped their fruits early, while others of their own kind continued to hold them. These circumstances helped to explain why some of the plants listed did not *always* bear their fruits during the seasons shown for them here. Generally they will.

You will note that red cedar and linden viburnum are listed as having berries on them during all four seasons of the year. This is true, except in the latter part of spring, when many of the old berries drop off, and in early summer, when a new crop of berries is forming.

### WINTER

| (December, January, and February) | Kind of Fruit and Its Color |
|---|---|
| Red cedar | Blue berries |
| Linden viburnum | Red berries |
| European cranberry-bush | Red berries |
| Japanese barberry | Red berries |
| Washington hawthorn | Small red applelike fruits |
| Sargent crabapple | Small dark-red apples |
| Small-leaved cotoneaster | Red berries |
| Box elder (a maple) | Seeds |
| Gray birch | Seeds |
| Amur honeysuckle | Red berries |
| Black haw | Blue-black berries |
| Virginia creeper | Blue-black berries |

*Winter:* total of twelve kinds

### SPRING

| (March, April, and May) | Kind of Fruit and Its Color |
|---|---|
| Red cedar | Blue berries |
| Linden viburnum | Red berries |
| European cranberry-bush | Red berries |
| Japanese barberry | Red berries |
| Washington hawthorn | Small red applelike fruits |
| Sargent crabapple | Small dark-red apples |
| Small-leaved cotoneaster | Red berries |
| Box elder (a maple) | Seeds |
| American elm | Buds, flowers, and seeds |

*Spring:* total of nine kinds

### SUMMER

| (June, July, and August) | Kind of Fruit and Its Color |
|---|---|
| Red cedar | Blue berries |
| Linden viburnum | Red berries |
| Russian mulberry | White to dark-red berries |

| | |
|---|---|
| Serviceberry (shadbush) | Small red applelike fruits |
| Raspberries and blackberries | Black berries and red berries |
| Blueberries | Blue berries |
| Fire, or pin, cherry | Small red cherries |
| Doublefile viburnum | Red berries |
| Seibold's viburnum | Red berries |
| White pine | Seeds |
| Gray dogwood | White berries |
| Japanese dogwood | Red raspberrylike fruits |
| Trumpet creeper | Flower nectar for hummingbirds |
| Common elderberry | Blue to purple-black berries |
| Arrowwood | Blue berries |
| Japanese yews | Red berries |
| Virginia creeper | Blue-black berries |

*Summer:* total of seventeen kinds

### FALL

| (September, Oct.ober, and November) | Kind of Fruit and Its Color |
|---|---|
| Red cedar | Blue berries |
| Linden viburnum | Red berries |
| European cranberry-bush | Red berries |
| Japanese barberry | Red berries |
| Washington hawthorn | Small dark-red fruits |
| Sargent crabapple | Small dark-red apples |
| Small-leaved cotoneaster | Red berries |
| Box elder (a maple) | Seeds |
| Gray birch | Seeds |
| Amur honeysuckle | Red berries |
| Black haw | Blue-black berries |
| Virginia creeper | Blue-black berries |
| Fire, or pin, cherry | Small red cherries |
| Gray dogwood | White berries |
| Japanese yews | Red berries |
| White pine | Seeds |
| Red pine | Seeds |
| White oak | Acorns |
| Flowering dogwood | Red berries |
| Orange-fruited viburnum | Orange berries |

*Fall:* total of twenty kinds

### SOME BIRDS OF THE UNITED STATES AND CERTAIN PLANTS ON THE FOLLOWING LISTS THAT ESPECIALLY ATTRACT THEM

| Kind of Bird | Plant Group |
|---|---|
| Bluebirds | Blackberries, wild cherries, dogwoods, wild grapes, cedars or junipers, mulberries, Virginia creeper, sumacs, blueberries, elderberries, serviceberries, hollies; also bayberries, hackberries, Russian olives, California pepper tree. |
| Cardinal | Blackberries, wild cherries, dogwoods, wild grapes, mulberries, sumacs, blueberries, elderberries, tulip tree, greenbriers; also hackberries and Russian olives. |
| Catbird | Blackberries, wild cherries, dogwoods, wild grapes, cedars or junipers, mulberries, Virginia creeper, sumacs, blueberries, elderberries, serviceberries, greenbriers, hollies; also bayberries, hackberries, persimmons, Russian olives, buffaloberries, buckthorns. |
| Chickadees | Pines, oaks, maples, spruces, Virginia creeper, blueberries, birches, elms, hemlocks, serviceberries, firs; also bayberries and sweet gum. |
| Crossbills | Pines, cedars or junipers, spruces, hemlocks, firs. |
| Duck (wood) | Oaks, beech, elms, greenbriers, hickories, ashes. |
| Finch (purple) | Dogwoods, cedars or junipers, maples, sumacs, black gum, elms, tulip tree, ashes, aspens; also sweet gum. |
| Flickers | (see woodpeckers) |

| | |
|---|---|
| Flycatcher (crested) | Wild cherries, dogwoods, mulberries, Virginia creeper. |
| Goldfinches | Maples, elms, tulip tree, alders; also sweet gum. |
| Grosbeaks | Pines, blackberries, wild cherries, dogwoods, oaks, cedars or junipers, maples, mulberries, spruces, sumacs, elderberries, beech, elms, serviceberries, hickories, ashes, mountain ashes, hawthorns; also hackberries, Russian olives, manzanitas, buffaloberries, snowberries. |
| Jays | Pines, blackberries, oaks, wild grapes, cedars or junipers, sumacs, blueberries, elderberries, beech, serviceberries, hickories; also bayberries and manzanitas. |
| Juncos | Pines; also sweet gum and Russian olives. |
| Kingbirds | Wild cherries, dogwoods, wild grapes, mulberries, blueberries, elderberries. |
| Mockingbird | Blackberries, dogwoods, wild grapes, cedars or junipers, mulberries, Virginia creeper, sumacs, black gum, elderberries, serviceberries, greenbriers, hollies; also hackberries, palmettos, persimmons, manzanitas, California pepper tree. |
| Nuthatches | Pines, oaks, maples, spruces, Virginia creeper, elderberries, beech, hickories, firs. |
| Orioles | Blackberries, mulberries, blueberries, elderberries, serviceberries. |
| Phainopepla | Wild cherries, wild grapes, mulberries, elderberries; also buckthorns and California pepper tree. |
| Phoebes | Blackberries, sumacs, blueberries, elderberries; also hackberries. |
| Quails | Pines, oaks, cedars or junipers, sumacs; also hackberries, prickly pears, wild roses, Russian olives, manzanitas, mesquite, buffaloberries. |
| Redpolls | Birches, alders. |
| Robin | Blackberries, wild cherries, dogwoods, wild grapes, cedars or junipers, mulberries, Virginia creeper, sumacs, blueberries, black gum, serviceberries, greenbriers, hollies; also hackberries, palmettos, persimmons, Russian olives, buckthorns, California pepper tree. |
| Siskin (pine) | Pines, spruces, birches, hemlocks, alders. |
| Solitaire | Wild cherries, cedars or junipers, sumacs, hawthorns; also hackberries and wild roses. |
| Sparrow (fox) | Blackberries, wild grapes, Virginia creeper, birches, greenbriers, hawthorns; also hackberries and manzanitas. |
| Sparrow (song) | Blackberries, wild cherries, elderberries. |
| Sparrow (tree) | Blueberries, birches. |
| Sparrow (white-throated) | Blackberries, wild cherries, wild grapes, blueberries, elderberries, greenbriers; also sweet gum. |
| Starling | Wild cherries, dogwoods, mulberries, sumacs, black gum, elderberries; also bayberries and Russian olives. |
| Swallow (tree) | Dogwoods, cedars or junipers, Virginia creeper; also bayberries. |
| Tanagers | Blackberries, wild cherries, dogwoods, wild grapes, mulberries, blueberries, black gum, elderberries, serviceberries; also bayberries and Russian olives. |
| Thrashers | Pines, blackberries, wild cherries, dogwoods, oaks, wild grapes, mulberries, Virginia creeper, sumacs, blueberries, black gum, elderberries, serviceberries, hollies; also hackberries, prickly pears, buffaloberries, buckthorns. |

| | |
|---|---|
| Thrushes | Blackberries, wild cherries, dogwoods, wild grapes, cedars or junipers, mulberries, Virginia creeper, sumacs, blueberries, black gum, elderberries, serviceberries, greenbriers, hollies, mountain ashes; also hackberries, wild roses, buckthorns, snowberries, Pacific madrone, California pepper tree. |
| Titmice | Pines, blackberries, oaks, wild grapes, Virginia creeper, beech; also hackberries. |
| Towhees | Pines, blackberries, wild cherries, oaks, blueberries, elderberries, serviceberries, hollies, aspens; also hackberries, sweet gum, Russian olives. |
| Vireos | Blackberries, wild cherries, dogwoods, wild grapes, Virginia creeper, sumacs, elderberries, hollies; also bayberries, wild roses, snowberries. |
| Warblers | Pines, blackberries, dogwoods, wild grapes, cedars or junipers, mulberries, sumacs, elderberries; also bayberries, persimmons, California pepper tree. |
| Waxwing (cedar) | Blackberries, wild cherries, dogwoods, wild grapes, cedars or junipers, mulberries, spruces, black gum, elder(cedar)berries, serviceberries, greenbriers, hollies, mountain ashes, hawthorns; also hackberries, persimmons, Russian olives, buffaloberries, California pepper tree. |
| Woodpeckers | Pines, blackberries, wild cherries, dogwoods, oaks, wild grapes, mulberries, spruces, Virginia creeper, sumacs, black gum, elderberries, beech, serviceberries, greenbriers, hickories, hollies, hawthorns, hazelnuts; also bayberries, hackberries, palmettos, prickly pears, buckthorns, California pepper tree. |
| Wren (cactus) | Sumacs, elderberries; also prickly pears. |
| Wren (Carolina) | Pines; also bayberries and sweet gum. |

## SOME RECOMMENDED TREES, SHRUBS, AND VINES FOR YOUR REGION

Listed in order of their value, from most important to least important, as food and shelter to birds.

### NORTHEAST REGION

Connecticut, Delaware, Indiana, Kentucky, Maine, Maryland, Massachusetts, Michigan, Minnesota, New Hampshire, New Jersey, New York, Ohio, Pennsylvania, Rhode Island, Vermont, Virginia, West Virginia, Wisconsin.

**Pine** Jack pine, Japanese black pine, Korean pine, red pine, pitch pine, white pine.

**Blackberry** Cultivated varieties 'Eldorado,' 'Snyder,' 'Taylor,' and 'Erie,' and recommended varieties for your area.

**Wild cherry** Sour cherry (not, strictly speaking, a "wild" cherry, but hardy and suitable for shady places, with fruit very attractive to songbirds), fire or pin cherry, wild black (or rum) cherry, chokecherry.

**Dogwood** Siberian dogwood (variety *sibirica*), pagoda dogwood, rough-leaved dogwood, flowering dogwood, Japanese dogwood, gray dogwood.

**Oak** White oak, red oak, scarlet oak, burr (or mossy-cup) oak, chestnut oak, pin oak.

**Wild grape** Summer grape, frost grape, fox grape, riverbank grape; also cultivated grapes.

**Cedar or juniper** Eastern red cedar, *virginiana,* and varieties *canaerti, keteleeri,* and *tripartita;* you must plant both male and female trees to produce berries for birds.

**Maple** Amur maple, red maple, sugar maple, and box elder; you may need to plant both male and female box elders to produce seeds for birds.

**Mulberry** White mulberry, Russian mulberry, red mulberry, and cultivated mulberry varieties 'New American,' 'Thorbum,' and 'Trowbridge'; you may need to plant both male and female mulberries to produce fruit for birds.

**Spruce** (*Picea.*)

**Virginia creeper** (*Parthenocissus quinquefolia.*)

**Sumac** (*Rhus.*)

**Blueberry** Highbush blueberry, lowbush blueberry; cultivated varieties (early) 'Cabot,' 'June,' 'Weymouth'; (midseason) 'Rancocas,' 'Concord,' 'Pioneer,' and 'Stanley'; (late) 'Atlantic,' 'Burlington,' 'Jersey,' 'Pemberton,' 'Rubel,' and 'Wareham'; and recommended varieties for your area.

**Black gum** (*Nyssa sylvatica.*)

**Elderberry** American elder, red elder.

**Beech** (*Fagus.*)

**Birch** Sweet or black birch, paper birch, gray birch.

**Elm** (*Ulmus.)*

**Tulip tree** (*Liriodendron.*)

**Hemlock** (*Tsuga.*)

**Alder** (*Alnus.*)

**Serviceberry** Downy serviceberry, apple serviceberry, Allegany serviceberry, thicket serviceberry, dwarf juneberry.

**Greenbrier** (*Smilax.*)

**Hickory** (*Carya.*)

**Ash** (*Fraxinus.*)

**Fir** (*Abies.*)

**Holly** Inkberry, American holly, long-stalk holly, winterberry; you may need to plant both male and female hollies to produce fruit for birds.

**Mountain Ash** (*Sorbus.*)

**Hawthorn** Cockspur thorn, Lavalle hawthorn, Washington hawthorn.

**Hazelnut** (*Corylus.*)

**Aspen** (*Populus.*)

## SOUTHEAST REGION

Alabama, Arkansas, Florida, Georgia, Louisiana, Mississippi, North Carolina, southeastern Oklahoma, South Carolina, Tennessee, eastern Texas.

**Pine** Slash pine, shortleaf pine, longleaf pine, loblolly pine.

**Oak** Laurel oak, blackjack oak, water oak, willow oak, live oak.

**Blackberry** Cultivated varieties 'Nanticoke,' 'Rathbun,' 'Agawan,' 'Dorchester,' 'Early King,' 'Lawton'; and recommended varieties of blackberries and raspberries for your area.

**Wild grape** Summer grape, frost grape, muscadine grape, currant grape; and cultivated varieties.

**Virginia creeper** (*Parthenocissus quinquefolia.*)

**Poison ivy** Although songbirds are fond of the white, berrylike fruits of these plants, poison ivies are not recommended for planting.

**Wild cherry** Wild black cherry, chokecherry; also cultivated cherries.

**Holly** Dahoon holly, possumhaw, American holly, winterberry, yaupon; you may need to plant both male and female hollies to produce fruit for birds.

**Greenbrier** (*Smilax.*)

**Waxmyrtle or Bayberry** Waxmyrtle, bayberry, dwarf waxmyrtle; you will need to plant both male and female bushes to produce fruit for birds.

**Blueberry** Cultivated varieties (early) 'Wolcott' and 'Murphy'; (midseason) 'Rancocas'; (late) 'Jersey'; and recommended varieties for your area.

**Dogwood** Siberian dogwood (variety *sibirica*), pagoda dogwood, rough-leaved dogwood, flowering dogwood, gray dogwood.

**Mulberry** Black mulberry, and cultivated variety 'Black Persian'; red mulberry, and cultivated varieties 'Hicks' and 'Stubbs'; Texas mulberry; you may need to plant both male and female mulberries to produce fruit for birds.

**Black gum** (*Nyssa.*)

**Beech** (*Fagus.*)

**Elderberry** American elder and its varieties.

**Cedar or juniper** Eastern red cedar, *virginiana* (and varieties *canaerti, ketaleeri,* and *tripartita*); you must plant both male and female cedars to produce berries for birds.

**Maple** (*Acer.*)

**Tulip tree** (*Liriodendron.*)

**Hickory** (*Carya.*)

**Ash** (*Fraxinus.*)

**Hackberry** Sugar or southern hackberry, and common hackberry.

**Palmetto** (*Sabal.*)

**Persimmon** (*Diospyros.*)

**Sweet gum** (*Liquidambar.*)

## PLAINS AND PRAIRIE REGION

Eastern Colorado, Illinois, Iowa, Kansas, eastern Montana, Missouri, Nebraska, North Dakota, Oklahoma, South Dakota, central and western Texas, eastern Wyoming.

**Hackberry** Common hackberry, sugar hackberry (or sugarberry).

**Wild grape** Long's grape, riverbank grape; also domestic or cultivated grape varieties recommended for your area.

**Wild cherry** Sour cherry (not, strictly speaking, a "wild" cherry, but hardy and suitable for shady places, with fruit very attractive to songbirds), western chokecherry (variety *dernissa*), fire (or pin) cherry, wild black cherry.

**Poison ivy** Although songbirds are fond of the white, berrylike fruits of these plants, poison ivies are not recommended for planting.

**Dogwood** Siberian dogwood (variety *sibirica),* pagoda dogwood, rough-leaved dogwood, round-leaved dogwood, gray dogwood.

**Holly** Possum-haw, inkberry, long-stalk holly, winterberry; you will need to plant both male and female hollies to produce fruit for birds.

**Prickly pear** (*Opuntia.*)

**Alder** (*Alnus.*)

**Mulberry** White mulberry and its cultivated varieties; also Russian mulberry; you will need to plant both male and female mulberries to produce fruit for birds.

**Serviceberry or saskatoon** Alder-leaved serviceberry, downy serviceberry, apple serviceberry, Allegany serviceberry, thicket serviceberry.

**Pine** Lodgepole pine (variety *latifolia*), limber pine, western yellow pine, or ponderosa pine.

**Oak** Shingle oak, burr oak, blackjack oak, post oak.

**Blackberry** Recommended varieties of blackberries and raspberries for your area.

**Sumac** (*Rhus.*)

**Wild rose** (*Rosa.*)

**Cedar** Creeping juniper, Rocky Mountain red cedar and its horticultural varieties; also eastern red cedar (and varieties *virginiana, canaerti, keteleeri,* and *tripartita*); you must plant both male and female cedars to produce berries for birds.

## MOUNTAIN AND DESERT REGION

Arizona, western Colorado, Idaho, western Montana, Nevada, New Mexico, eastern Oregon, western Texas, Utah, eastern Washington, western Wyoming.

**Pine** Mexican pinon, pinyon (variety *edulis,* also called nut pine), limber pine (also called Rocky Mountain white pine), shore pine, lodgepole pine (variety *latifolia*), western yellow pine, or ponderosa pine.

**Serviceberry or saskatoon** Alder-leaved serviceberry, Baker serviceberry, Cusick serviceberry, Utah serviceberry.

**Hackberry** Douglas hackberry, common hackberry, thick-leaved hackberry.

**Cedar or juniper** Cherrystone juniper, western juniper,

Rocky Mountain red cedar and its horticultural varieties, Utah juniper; you must plant both male and female cedars to produce berries for birds.

**Russian olive** Russian olive, cherry elaeagnus, and autumn elaeagnus.

**Alder** White alder, mountain (or thin-leaved) alder.

**Gooseberry** Not recommended for planting because both gooseberry and currant plants are intermediate hosts for white pine blister rust, a deadly fungus that kills five-needled (or white) pine trees.

**Oak** Canyon live oak, Emory oak, Gambel oak, gray oak, Rocky Mountain white oak.

**Manzanita** Great-berried manzanita, greenleaf manzanita, pointleaf manzanita, bearberry.

**Blackberry** Recommended varieties of blackberries and raspberries for your area.

**Mesquite** Mesquite (*Prosopis glandulosa*).

**Wild grape** Canyon grape, California grape; also cultivated grape varieties recommended for your area.

**Prickly Pear** (*Opuntia*.)

**Wild cherry** Bitter cherry, fire (or pin) cherry, western chokecherry (variety *demissa*).

**Birch** Red (or water) birch.

**Fir** White fir, Alpine fir, corkbark fir (variety *arizonica*).

**Spruce** Engelmann spruce, white spruce, Alberta spruce (variety *albertiana*), blue spruce and its varieties.

**Buffaloberry** Silver buffaloberry, Canadian buffaloberry; you must plant both male and female buffaloberries to produce berries for birds.

## PACIFIC REGION

California, western Oregon, western Washington.

**Pine** Shore pine, Coulter pine, Jeffrey pine, sugar pine, western white pine, western yellow pine (or ponderosa pine), digger pine.

**Oak** Coast live oak, canyon live oak (or golden-cup oak), California blue oak, Oregon white oak, California black oak, valley white oak.

**Poison oak** (*Rhus diversiloba,* not recommended for planting.)

**Elderberry** Blueberry elder, redberry elder, black elder.

**Blackberry** Recommended varieties of blackberries, dewberries, loganberries, etc., for your area.

**Buckthorn** Alder-leaved buckthorn, California buckthorn (or coffee berry), hollyleaf buckthorn (or red berry), cascara buckthorn; some buckthorns, particularly hollyleaf (*Rhamnus crocea*), are secondary hosts for the crown rust of oats, and should not be planted near oats.

**Wild cherry** Bitter cherry, hollyleaf cherry, Catalina cherry, western chokecherry (variety *demissa*).

**Dogwood** Creek dogwood, brown dogwood, Pacific dogwood, western dogwood.

**Snowberry** Snowberry, mountain snowberry, round-leaved snowberry.

**Alder** White alder, red alder, Sitka alder, mountain (or thin-leaved) alder.

**Mistletoe** (*Phoradendron* and *Arceuthobium,* green plants parasitic on trees, not recommended to establish—dwarf mistletoes are harmful to evergreens, especially in the West.)

**Cedar or juniper** California juniper, western juniper, Rocky Mountain red cedar and its horticultural varieties; you must plant both male and female cedars to produce berries for birds.

**Manzanita** Parry manzanita, greenleaf manzanita, pointleaf manzanita, Stanford manzanita, woolly manzanita, bearberry.

**Gooseberry** (*Ribes*, not recommended for planting because both gooseberry and currant plants are intermediate hosts for white pine blister rust, a deadly fungus that kills five-needled, or white, pine trees.)

**Pacific madrone** (*Arbutus menziesi*.)

**Serviceberry** Alder-leaved serviceberry, western serviceberry.

**Prickly Pear** (*Opuntia*.)

**Birch** Red (or water) birch.

**Fir** White fir, lowland white fir (or giant fir), Alpine fir, California red fir, noble fir.

**Spruce** Engelmann spruce, Alberta spruce (variety *albertiana*), Sitka spruce.

## SOME GARDEN FLOWERS WHOSE SEEDS ATTRACT BIRDS

These attract the seed-eating songbirds, especially those that either normally or occasionally feed on or near the ground. Some of them include doves, quails, blackbirds, buntings, cardinals, cowbirds, finches, grackles, grosbeaks, juncos, larks, native sparrows—the chipping, fox, white-throated, white-crowned, tree, and song sparrows—and towhees and thrashers. If you plant some of the flowers on the above list in your garden and let them go to seed, they will provide summer, fall, and winter foods for birds.

Amaranthus, love-lies-bleeding (*Amaranthus caudatus*), and prince's feather (*Amaranthus hybridus* var. *hypochondriacus*) are especially good.

Asters, especially China aster (*Callistephus chinensis*), which has full heads of seeds.

Batchelor's buttons, or cornflower, (*Centaurea cyanus*).

Calendula.

Campanula, or bellflower.

Carduus, especially blessed thistle (*Carduus benedictus*).

Chrysanthemums.

Columbines (*Aquilegia*).

Coneflowers (*Rudbeckia*), or black-eyed Susans.

Coreopsis, or tickseed.

Cosmos.

Dianthus (sweet Williams and pinks).

Forget-me-nots (*Myosotis*).

Four-o'-clocks (*Mirabilis*).

Gaillardias.

Larkspurs, or delphiniums.

Marigolds (*Tagetes*).

Petunias.

Poppies, particularly *Eschscholzia californica*, the California poppy.

Portulaca, or purslane. (The flowers of these plants produce especially abundant crops of seeds.)

Phlox, particularly the annual, *Phlox drummondi*.

Scabiosa, mourning bride, or pincushion flower.

Sunflowers (*Helianthus*).

Tarweed, particularly *Madia elegans*, the common tarweed.

Verbena, especially the common garden verbena, *Verbena hybrida*.

Zinnias.

# Part 3

## *The Struggle to Survive*

# *How You Can Help the Birds*

ractically all wild animals can be "helped," or their populations increased, by improving their living conditions. Songbirds can be attracted in greater numbers and variety to your garden by providing those things that birds need that may be lacking in your yard. Increasing songbird foods, shelters, and nesting places will be remarkably effective.

## *How Much Cold Can Your Birds Endure?*

How can the tiny chickadees and juncos wintering in your garden live through nights when the wind roars through the trees and sweeps through your shrubbery like the cold, cutting edge of a knife? That many of them can survive at all is, to me, one of the most astonishing facts in all the out-of-doors. How do they do it?

Within the small feathered body of a chickadee, weighing less than an ounce, a small furnace with a tiny heart for a pump and a maze of blood vessels to carry heat keeps its body aglow with a warmth that will keep the bird alive at 30 and 40 degrees below zero. But this tiny furnace, like that of your own body, or the furnace down in your cellar, needs fuel; and fuel to a bird, or to any living creature, is food.

In Ohio a few years ago, a scientist became interested in this problem. How long could small birds survive cold weather without food?

First, he experimented with house sparrows. In our northern states, where we usually have the coldest winters, the nights are fourteen to fifteen hours long. House sparrows could live, exposed to the cold through a night this long, *if the temperature went no lower than 5 degrees above zero*. At 20 degrees *below* zero they could

**COMMON GRACKLE**

| | |
|---|---|
| Number of Eggs Laid in Clutch: | 4-7 |
| Days to Hatch: | 14 |
| Days Young in Nest: | 18-20 |
| Number of Broods Each Year: | 1 |
| Lifespan: | 4 to 16 years |

*Grackles nest in the summer throughout most of North America east of the Rockies. They build nests of twigs, weeds, and rags in tall evergreens, elms, maples, in the low bushes of swamps and marshes, and in the lower parts of osprey nests.*

live only ten hours, or about two-thirds of the night; at 30 degrees below zero, they lived only seven hours, or about half the night.

How then can chickadees, juncos, and tree sparrows—which he discovered will live through slightly colder weather than house sparrows—how can they live through a night when the temperature drops to 30 or 40 below zero?

The answer is: food and shelter. If they have eaten well late in the day, or just before going to bed, and if they are able to find an evergreen thicket, or other sheltered nook in which to sleep, then their own body heat carries them through the cold winter night.

## The Daily Requirements of Garden Birds

The daily requirements of birds are about the same as your own. They need food, shelter, and water. Give them the proper food and you may attract some birds to your yard, but give them food in the right relation to shelter and more will come. Provide them with bathing and drinking water, besides food and shelter, and soon, if you don't live in the middle of a city, your yard will be so enlivened by birds that it will look like an aviary.

RULE OF THUMB

Remember that the three essential requirements of birds are food, shelter, and water.

Shelter to a bird does not mean a roof over its head. Shelter means a bush or tree into which the bird can fly to safety. The tree or bush may also offer a place to build a nest in summer, or it may provide some kind of seasonal food or protection from the cold winds and snowstorms of winter. Supply these requirements in your yard and you have built for this bird, and for others, both *a dining room and a bedroom*.

Birds are certain to come to you then; for, quite simply, the secret of attracting them is in knowing what their needs are and giving them what they require. Once you try this, you'll be surprised how quickly they will move in on you.

# The Struggle to Survive—The Bird's Year

Perhaps we should take the time to summarize the natural changes during each season that affect the lives of birds throughout the year. The summary that follows applies specifically to birds and seasonal changes in the northern half of the United States, across the continent, and south to wherever winter brings freezing weather, ice, or snow. Even so, winter in the South, where it is usually mild, can be critical for birds if they are short of natural food.

## What Winter Means to the Birds

In many parts of the United States, winter means spells of bitterly cold weather. Snow and ice often cover seeds and berries that have fallen to the ground. Many insects have been killed by the cold, retreated underground, or hidden inside trees and in other places where they sleep through the winter.

Leaves have fallen from most plants and there is a general scarcity of protective cover where birds can roost at night and find shelter from cold winds, rain, and snow. In our garden, we determined that what winter birds needed was as follows:

*Food:* Fleshy fruits of barberries, hollies, cedars, mountain ashes, crabapples, and hawthorns that persist on the plants into spring are valuable emergency foods. Also important are the seeds of maples, ashes, birches, alders and the seeds of pines, spruces, hemlocks, and other evergreens, as well as the acorns of oak trees. Your feeding stations will help.

*Shelter:* Especially important are pines, hemlocks, spruces, yews, and firs where birds can roost at night or escape from cold winds and storms in daytime. Birdhouses and roosting boxes will help, as will plantings in your yard and garden.

## Old and New Names for Birds

In 1973, a committee of the American Ornithologists' Union published a revised list of some North American bird names, both common and scientific. These changes were not made capriciously, or arbitrarily, but were based on scientific studies that have shown, for example, that the so-called yellow-shafted flicker of the eastern United States and the red-shafted and gilded flickers of the western United States, which hybridize extensively where their ranges overlap, are not three distinct species, but one. The name common flicker has now been given to all three birds. Similarly, the Baltimore oriole of the eastern United States and the Bullock's oriole of the West, which hybridize freely where their ranges meet, are considered to be one species with the now common English name of northern oriole. Catbird is now "gray" catbird specifically for its gray color; American goldfinch and American robin now distinguish these two from the European goldfinch and European robin, and so on.

Because one finds in the older bird books and articles only the older names, I have listed below the old and new names and the family to which each bird belongs. Name changes follow those of the *Check-list of North American Birds* (American Ornithologists' Union, sixth edition, 1983).

| OLD NAME | PREFERRED NEW NAME | FAMILY | SOME LOCAL OR OLD AND LESS COMMON NAMES |
|---|---|---|---|
| Catbird | Catbird, gray | Mockingbird | Black mockingbird, black-capped thrush |
| Flicker, yellow-shafted | Flicker, common | Woodpecker | High hole, yellow-hammer, wicker (from its calls), golden-winged woodpecker |
| Flicker, red-shafted | Flicker, common | Woodpecker | Red-hammer, yucker (from its calls) |
| Flicker, gilded | Flicker, common | Woodpecker | Wa-cup, or wake-up (from its calls) |
| Goldfinch, common | Goldfinch, American | Finch | Lettuce-bird (eats leaves or seeds of), salad bird, thistle-bird (eats seeds of thistle), wild canary, yellow bird |
| Grackle, purple or bronzed | Grackle, common | Troupial (blackbirds and orioles) | Crow, blackbird, maize thief |

| OLD NAME | PREFERRED NEW NAME | FAMILY | SOME LOCAL OR OLD AND LESS COMMON NAMES |
|---|---|---|---|
| Hawk, sparrow | Kestrel, American | Falcon | Grasshopper hawk, killy hawk, windhover |
| Jay, Florida | Jay, scrub | Crow (jays, crows, ravens, magpies, nutcrackers) | Bullfinch, scrub jay (from habitat in Florida scrub), smoky jay (from color) |
| Jay, California | Jay, scrub | Crow | Long-tailed jay, Nicasio jay, Santa Cruz jay, Texas jay, Woodhouse's jay, Xantus' jay |
| Junco, Oregon | Junco, dark-eyed | Finch | Mountain junco, pink-sided junco |
| Junco, slate-colored | Junco, dark-eyed | Finch (buntings, cardinal, grosbeaks, finches, sparrows, towhees, longspurs) | Black chipping bird, black snowbird, Carolina junco, snowbird, common snowbird, white-bill |
| Junco, white-winged | Junco, dark-eyed | Finch | Snowbird |
| Oriole, Baltimore | Oriole, northern | Troupial (blackbirds and orioles) | Firebird, golden robin, hang-nest |
| Oriole, Bullock's | Oriole, northern | Troupial | Firebird, hang-nest |
| Robin | Robin, American | Thrush | Common robin, robin redbreast |

*Water:* Most streams and ponds periodically freeze over, so keep the water in your birdbath thawed and supply the bath with fresh water daily.

## WHAT SPRING MEANS TO THE BIRDS

Times get better for birds in spring. Snow and ice start to disappear in March and April, uncovering the seeds and fallen fruits of the previous year. Insects hatch from eggs or awaken from winter sleep as the days grow warmer. New leaves open and, day by day, offer more protective and nesting cover for birds.

The danger remains, however, of early spring snowstorms and late cold weather, or of prolonged spring rains—hardships that make this period still a critical one for many kinds of birds. In general, their needs are as follows:

*Food:* As in the winter, the fleshy fruits of barberries, hollies, cedars, mountain ashes, crabapples, and hawthorns are all important. These provide valuable emergency foods, as do the seeds of maples, ashes, birches, alders, seeds of pines, spruces, hemlocks, and other evergreens, and the acorns of oak trees. Again, your

### Petting a Wild Bird

Mabel T. Tilton of Vineyard Haven, Massachusetts, got a seemingly fearless red-breasted nuthatch to come to her to feed in winter. While on her hand, it sometimes picked at her fingernails and seemed to enjoy warming its cold feet in her hand. It was never in haste to leave. Once, when she put her thumb over its short tail, it was not disturbed but merely turned away from her and looked all about from the security of her hand.

### A Remarkable Memory for Hidden Stores of Food

Some members of the crow family show a remarkable "place memory" for hidden stores. Eurasian thick-billed nutcrackers relocate their caches of pine seeds even when they are covered by snow: They fly to the spot and dig directly down to them. A Russian scientist, in a series of experiments, discovered that these birds, which recover up to 70 percent of their stored foods, visually remember the exact places where they bury them. Victor Cahalane reported that in the Rocky Mountains a Clark's nutcracker dug directly down through eight inches of snow to recover a seed cone of a Douglas fir that it had apparently stored there.

feeding stations will help.

*Shelter:* The evergreens (pines, yews, spruces, etc.) are still valuable protection against spring cold and storms. Plant trees, shrubs, and vines in your yard at this time. Spring is the best time to plant your evergreens; it is also an excellent time to plant other trees, shrubs, and vines.

*Water:* Keep birdbath water thawed in early spring and supply fresh water daily.

*Nesting Places:* If you didn't do so the previous fall, this is the time to put up bird boxes—the earlier the better.

## What Summer Means to the Birds

Summer is probably the most "prosperous" time of the whole year for birds. Most of them raise their families in summer. Berries and insects to feed their youngsters and themselves are abundant. Cover—the leafy shelter of trees, shrubs, and vines—is more luxuriant than at any other time of the year.

Yet summer may still be a critical time for some birds, if they have no place to nest. If they do have nesting places, but must travel far to get fresh berries to feed their young ones or for drinking and bathing water, they may be exposed to extra dangers. Food and water close by will eliminate or lessen these hazards. In general, here is how you can help them:

*Food:* Fleshy fruits of cherries, mulberries, serviceberries, blackberries, raspberries, blueberries, and elderberries are especially needed now by many nesting birds. You can help by having some of these plants in your yard.

*Shelter:* In summer, birds look for pines, spruces, yews, barberry hedges, grapevine thickets, hawthorns, and hollow trees. You can help by putting up bird boxes, and your spring-planted trees, shrubs, and vines will help too.

*Water:* This is more important now, especially in hot, dry weather, than at any other time of the year. In fact, it is more

**CHIPPING SPARROW**

| | |
|---|---|
| Number of Eggs Laid in Clutch: | 3-5 |
| Days to Hatch: | 11-14 |
| Days Young in Nest: | 8-12 |
| Number of Broods Each Year: | 1 or 2 |
| Lifespan: | 2 to 10 years |

*The chipping sparrow ranges from Alaska and Canada south throughout most of the United States. It often builds its nest of grasses, weeds, rootlets, and hair (off of cattle, horses, and people) on the limb of a garden spruce, pine, or cedar, or in orchard trees 1 to 40 feet above the ground (usually below six feet).*

**Preventing Birds from Flying into Picture Windows**

We don't have a picture window in our house, but a woman in North Carolina wrote to me about one in her home that accounted for the deaths of at least a dozen hummingbirds before she found a way to prevent it. The birds, seeing the reflected trees, shrubs, and grass in the window, apparently saw them as a further extension of the garden. When they tried to fly into it, they struck the plate-glass window and were killed. The woman bought some of the sheerest nylon marquisette she could get—enough to cover completely an area 6 feet long by 5 feet high, or the size of the picture window. The plate glass of her window was set back about two inches from the outside face of the house and window frame. She had the nylon stretched tight across the window opening and held in place by a narrow strip of molding tacked around the outside edge of the window frame. In this position the nylon screen prevents her hummingbirds from flying into the window yet does not obstruct her view when looking out because the nylon is sheer. She renews this inexpensive material every two or three years.

important to keep birdbaths filled with fresh water than to operate feeding stations during the summer.

## WHAT FALL MEANS TO THE BIRDS

In fall, food and cover gradually decline. Berries are eaten up by birds or fall to the ground. Leaves fall and the range of shelter for birds shrinks. Insects die from frosts or hibernate in the ground and in other places inaccessible to many birds. Weather gets colder and the days get shorter, which means that birds have less time to hunt for food than during the long summer days. Here is what birds especially need in fall:

*Food:* They look for the fleshy fruits and seeds that persist on plants above the snows of early winter. This is a good time to plant many kinds of trees, shrubs, and vines, excepting the "needle-evergreens" (pines, yews, etc.), which seem to do better if planted in spring rather than in fall.

*Shelter:* Pines, spruces, and other evergreens become increasingly important to birds, for shelter against the cold and storms of early winter. Bird nesting boxes and roosting boxes will help. Fall is the best time to put them up because the bird nesting boxes, if weathered, may be more acceptable to birds the next spring.

*Water:* Keep birdbaths thawed and fill them with fresh water daily.

## Birds That Store Food

Many people who feed birds have noted that, at times, white-breasted nuthatches, tufted titmice, and red-bellied and red-headed woodpeckers, instead of eating at the feeder, will snatch up a piece of suet, a sunflower seed, or bread and fly away with it in the bill. If near woods, they will shove the food into bark crevices or into knotholes and the crotches of trees or even into cracks in fence posts and utility poles. Some will tuck food under leaves or into soft ground. Other birds, however, may quickly steal the aboveground stores. John V. Dennis at his feeding station in Virginia noted that in one hour a male white-breasted nuthatch carried away thirty-eight pieces of suet, which it pushed into crevices of bark. Most of these stores were discovered a little later and eaten by a brown creeper, chickadees, and other nuthatches. Most birds do not defend their caches, but Lewis' and red-headed woodpeckers vigorously chase other birds from their winter stores of acorns and nuts, wedged into the bark of trees or the cracks of poles.

Food that birds hide under leaves or store in the ground, however, may not be stolen. Members of the crow family are great storers of food. Jays store food everywhere: We have watched the blue jays

**SONG SPARROW**

| | |
|---|---|
| Number of Eggs Laid in Clutch: | 3-6 |
| Days to Hatch: | 12-13 |
| Days Young in Nest: | 10 |
| Number of Broods Each Year: | 2, 3, or 4 |
| Lifespan: | 2 to 10 years |

*The song sparrow ranges across Canada and the United States, coast to coast, from Alaska south to Mexico. Early in the nesting season, song sparrows build most of their nests on the ground from grasses, weeds, bark, and animal hairs. Later, they nest 2 to 4 feet above the ground in a small conifer, shrub, in cattails, or in pine saplings.*

**HOUSE SPARROW**

| | |
|---|---|
| Number of Eggs Laid in Clutch: | 3-7 |
| Days to Hatch: | 11-12 |
| Days Young Are in Nest: | 15-17 |
| Number of Broods Each Year: | 3 or more |
| Lifespan: | 13 years |

*The English, or house, sparrow, first introduced successfully into the United States by Nicholas Pike in the spring of 1853 at Greenwood Cemetery, Brooklyn, New York, is not a true sparrow. It is a member of the Old World weaverbird family, called Ploceidae. Its nesting range extends across central and southern Canada, and throughout most of the United States, except for parts of the West.*

*The house sparrow will fill a birdhouse with grass, chicken feathers, cotton, and string. It also nests in the hollows of trees, on shutters of houses, and under eaves. The nests that this bird builds in trees are dome shaped.*

## Another Accident You Can Prevent

Birds, like ourselves, are creatures of habit. They get accustomed to flying in certain directions to and from their nests. Sometimes they follow aerial "paths," as we follow our trail of flagstones when we walk along the edges of our garden.

In the spring of 1982, because he left his garage doors open, one of our next-door neighbors lost one of the pair of catbirds that nested in his yard. Ordinarily he keeps the doors closed. On this day he had backed his car a little way down the driveway and had propped the doors wide open. The catbird, accustomed to flying along one side of the garage and then turning sharply to fly across the front of it to reach its nest, struck one of the doors and was killed.

A man I know near Philadelphia, who has been attracting birds and banding them for years, lost a male cardinal in the same way. Early in his bird-attracting and bird-banding career he had a pair of cardinals nesting in his yard. He had banded both birds. By their band numbers he knew them to be the same mated pair that had lived in his garden for five years. One day he, too, left open one of his garage doors, which was usually closed. The cardinal, flying swiftly along one of its regular aerial routes, turned the corner of the garage and struck the open door before it could check its speed or swerve to avoid it.

in our Little Neck garden pick up in their bills acorns and hazelnuts, sunflower seeds, and peanuts in the shell from our main bird feeder and fly with them to the ground under our shrub border. There, using the bill, they thrust them under leaves or into the soft earth. The western Steller's jay similarly buries acorns, and, in California, people have seen scrub jays burying pieces of bread, potato chips, grapes, peanuts, apple slices, pretzels, cheese crackers, cookies, hard-boiled eggs, sunflower seeds, and even marbles. Later, the buried sunflower seeds sprouted into ten-foot-tall stalks all over hillsides and gardens, but, in the southern California climate, never flowered and grew seed heads.

In our garden we have watched tufted titmice take sunflower seeds from our bird feeders and store them in crevices of bark and also tuck them into the grassy sod of our lawn, but we have never seen them return to dig up food they stored in the ground. In North Carolina, one February day we watched a tufted titmouse pick up a sunflower seed from our feeder and fly directly with it to a blooming camellia bush, where it tucked the seed deep inside one of the compact red flowers. Ten minutes later, either the same bird or another returned, picked up the seed, and flew away with it.

## Getting Acquainted with Birds

If all the work of building nesting boxes, planting trees and shrubs, and setting out feeders seems intimidating, remember that the first step to helping birds is simply getting acquainted with them and their habits. Getting acquainted with birds is pretty much like getting acquainted in your new neighborhood. You must learn what the Joneses look like before you can tell them from the Finneys or the Johnsons.

RULE OF THUMB

For beginners, house sparrows and starlings are two easily recognizable bird species.

If you have lived in the heart of any large city in our northeastern states, you have seen the little brown "English" sparrow, or house sparrow, introduced into this country more than a hundred years ago, and the dark-feathered European starling, a more recent arrival. These two are common birds that you will find in city backyards, in city streets, in shade trees, and perched on buildings almost everywhere. If you haven't noticed birds before, it will help you if you know the looks and actions of these two. Like the Joneses and the Finneys, the first two families that you meet in your new neighborhood, you'll soon know them well enough to distinguish them from your other neighbors. English sparrows and starlings are easy to identify, and knowing them will help you to recognize the differences between them and the native birds that are going to visit you. If you buy one of those books written especially to help you identify birds, it will be an inexpensive and worthwhile investment. I particularly recommend Roger Tory Peterson's excellent field guides or bird guides by Audubon's Richard Pough. There are other bird identification books listed in the Recommended References (Appendix 4) that may also be helpful to you as you begin.

TRY THIS

A field guide specific to your region of the country, especially one with color photographs, will help you identify the birds that come to your yard and garden.

When you have reached the point where you begin to notice birds and distinguish between them, you will probably fall under a spell as old as mankind—the *charm* of birds. You will make the exciting discovery that birds are, in many ways, like people—shy and timid, because of human persecution, yet easily won over by your kindness and consideration.

You will discover, above all, that birds are *individuals*, as different in their characters as you and I, which may account for the fascination they hold for men and women everywhere.

Many years ago, in the yard of my home in the country, I started to attract birds. I have done so ever since. Even after my work compelled me to move to the suburbs of New York City, hundreds of birds came to my home grounds because I provided for them. Besides helping protect my trees, shrubs, flowers, and garden crops from insects, I have found, as you will, that attracting birds is a lot of fun. It is healthful because you are much out-of-doors with your bird-attracting program, and it is fascinating and deeply satisfying. Anyone, whether young or elderly, can attract birds and help them survive, and you can enjoy them from your porch and yard, or, if you prefer, in the comfort of your house while sitting at the windows.

# The Care and Feeding of Young Birds

very spring and through the early summer Audubon House, headquarters of the National Audubon Society in New York City, gets a flood of telephone calls. When I was editor there, I took many of these calls, and the question most frequently asked was: "What should I feed a young bird that has fallen out of the nest?" Before these kindhearted people told me their stories, I could have told them what had happened.

RULE OF THUMB

Once feathered, a young robin walking about your lawn is in no danger—and does not require your assistance—as it can easily flutter up into the lower limbs of bushes or trees and out of reach of predators.

Someone in the household saw an apparently helpless half-feathered young bird on the lawn. The parent birds weren't in sight and it looked as if the young bird had been abandoned. So they brought it into the house to save it from starving or sudden death from the neighbor's cat. Perhaps these kindly people did rescue the young bird, but there was a better way to save it, with far less trouble to themselves.

When young birds leave the nest, they don't always fly away. Sometimes they "walk" from the nest and by a series of hops climb out into the bush or tree in which they were hatched. There they sit and call loudly; the parents, attracted by their cries, or "food calls," come to the youngsters at regular intervals to feed them. The young but feathered robins we see on the lawn, even if they can't fly strongly, are usually able to flutter up into the lower limbs of a bush and up into trees until they are safely out of the reach of dogs and cats. If you will remember this the next time you see a young robin on your lawn, and place it in a bush or tree out of reach of its enemies, you may save yourself a lot of trouble. You will also be sure that the young bird will be well fed with wireworms, white grubs, and cutworms from your garden, and earthworms from your lawn.

The next thing to do is call for help. Since *Songbirds in Your Garden* was first published, many networks have sprung up linking people trained and experienced at rehabilitating injured or stranded birds. In many parts of the country, a call to your local Humane

Society or animal shelter will get you a list of names and numbers to call. Another source for these numbers is your state wildlife office.

If you have already brought the young bird into the house, are unable to get help, and have accepted your responsibility of feeding and caring for the bird, here is what this will mean to you: You are now a foster parent and once you start feeding the youngster you will be astonished at how quickly he will learn what your relationship means to him. To the young bird, you have become his world, and a young bird's world is *food.*

## Accidents May Thrust Young Birds upon You

Because she did such an excellent job of raising young birds, I especially remember a Missouri woman who raised four baby robins a few years ago. Someone in her neighborhood cut down a tree in which a pair of robins had a nest filled with young ones. When the tree crashed to the ground, the recently hatched robins were thrown out of the nest and scattered over the lawn, but were unhurt.

The woman gathered up the four helpless creatures and put them in a sewing basket about 8 inches deep, which she had lined with excelsior to simulate the lining of grasses in their former nest. Then she wired the basket securely on top of a branch of a tree near the tree that had been cut down, hoping that the parents would return and feed the young ones.

Sometimes if a fallen nest and its young ones are replaced in the exact site from which the nest fell, or within a few feet of the original spot, the adult birds will return and continue to feed their young as though nothing had happened. But all birds, like all human beings, cannot be depended upon to behave in the way we think they should. If the adults are disturbed or frightened at the nest when the eggs are newly laid or the young just hatched, they

### Food for Young or Injured Birds

In feeding and caring for orphaned baby birds, give them plenty of canned dog food (meat), along with the chopped yolks of hard-boiled eggs, and mashed walnuts.

Young hummingbirds or injured adults in captivity should be fed the following (feed young hummingbirds with an eyedropper; allow adults to sip food from a small vial): To 4 ounces of water, add and stir until dissolved 1 level teaspoon of Mellon's food (or pabulum), ½ teaspoon of condensed milk, 1½ teaspoons of honey, 2 or 3 drops of Vipenti (a vitamin), and 1 drop of beef extract.

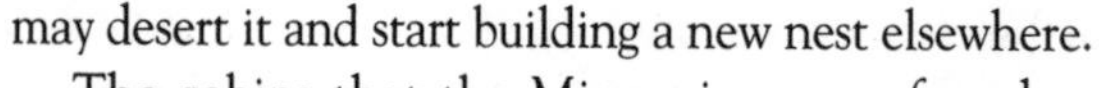

RULE OF THUMB

Adult birds will often return and continue to feed their young if the fallen nest is placed within a few feet of the exact site from which it fell.

may desert it and start building a new nest elsewhere.

The robins that the Missouri woman found were extremely young, scarcely more than a day old. Keeping some distance away, so as not to alarm the parent birds if they came back, she watched for a while over her improvised basket-nest. When the adults did not return, she realized they had abandoned their family. She was frightened at the prospect of raising the tiny young birds, but determined that she would not allow them to die without trying to save them.

## Feeding the Young Robins

Knowing that robins sometimes feed themselves and their young ones on earthworms, she got her two children to dig in the garden until they had a large jar filled with them. With a pair of old sugar tongs, she dropped small pieces of the earthworms in the robins' mouths, which they opened wide whenever they were hungry. Meanwhile, she had cooked some oatmeal. When the oatmeal had cooled, she fed them some of it, along with small pieces of whole wheat bread, moistened slightly with milk. To this diet she added pieces of scraped apple and small bits of cherries. Regularly, at least once every half hour, she fed the young robins that first day, and at night she covered the basket with a cloth to keep the birds warm.

WARNING

You are required by state and federal laws to release songbirds once they have proven capable of caring for themselves.

As the birds grew older, their appetites grew larger, and she increased her feedings to once every fifteen minutes. Within two weeks the young robins had feathered out and were so active that they would no longer stay in the basket. Then the woman had a chicken-wire cage, 10 feet long and 3 feet high, built for them in her backyard.

For a week the young birds lived there, flying from one branch to another of the limbs of trees she had put in the cage for them. They were now learning to hunt for worms and to pick up insects from the ground. The woman had become deeply attached to her four young birds, but legally she had no right to keep them once they had proved they were able to care for themselves. All songbirds are protected by federal and state laws, except for house sparrows and starlings.

When her robins were strong enough to fly, the woman allowed them to go free in the yard. For a few days they returned at her call to be fed pieces of bread and ripe blackberries. One day they flew off to a nearby woodland and she never saw them again. Presumably they survived and, though she regretted to see them go, she had acted commendably and sensibly. She had granted them their natural right to live the rest of their days wild and free.

**BULLOCK'S (NORTHERN) ORIOLE**

| | |
|---|---|
| Number of Eggs Laid in Clutch: | 3-6 |
| Days to Hatch: | 14 |
| Days Young in Nest: | 14 |
| Number of Broods Each Year: | 1 |
| Lifespan: | 6 to 7 years |

*The Bullock's oriole is common throughout most of the West, from the Great Plains to California. It nests in an oval woven bag, 6 inches deep, attached to the outer branches of cottonwood, birch, and the shade trees about ranches. The Bullock's oriole is known to hybridize with the Baltimore oriole in the Great Plains where the ranges of these two birds meet.*

## *The Attachment of Some Birds to Their Eggs and Young*

Not all birds will desert their young or their eggs when they are disturbed on the nest. I have stroked the backs of red-eyed vireos and chestnut-sided warblers as they sat on their nests, incubating their eggs or brooding their young ones to keep them warm.

One of the most remarkable examples of a parent bird's devotion to its offspring came to me from a man in Michigan. A pair of northern orioles built their nest one summer in an apple tree in his yard. When a cat killed the female, the male took over the feeding of the young. One day this bird broke its wing and when the man discovered that the oriole was crippled, he saw that it was still carrying food to its youngsters!

A grapevine had grown up into the lowest branches of the apple tree to a point where a pole leaned at an angle against the limb where the nest was hung. This formed a continuous pathway from the ground to the nest and up and down this "road" the poor bird searched for food. He never went far from the grapevine and kept a keen watch for enemies. After filling his beak with caterpillars and

**Birds That May Nest in Your Garage**

Robins, phoebes, or barn swallows may build their nests on the rafters and inside ledges of your garage if you leave the door open for a few days in spring or early summer. If you want the birds to nest in your garage, you should leave the doors open all through the nesting season so they can get in and out to feed their young ones or to incubate their eggs. Hungry young birds, shut off from their parents, may starve to death within a day if deprived of food. If you don't want birds to nest in your garage, be sure to keep the doors closed in spring and early summer to prevent a nesting tragedy.

RULE OF THUMB

For extremely young songbirds, make a nest of grass, excelsior, or torn-up paper in a shoe box.

other insects, he started his tedious journey up the grapevine—one hop at a time. In this way he cared for his family until the birds were able to fly and care for themselves. The crippled oriole disappeared soon after that, perhaps a victim of the same cat that had killed his mate.

## *The Substitute Nest for the Young Bird*

After you have raised quite a number of young birds, you get a local reputation for being a "bird doctor." People believe you can perform all sorts of miraculous cures. In the spring, our neighbors have brought us many kinds to be helped—from hour-old naked sparrows, with their eyes not yet open, to sick and crippled adult jays and crows. We always try to help these unfortunate creatures. If we have so many in our care that we can't possibly take on more, we are glad to tell our friends how they may care for them until these waifs are able to shift for themselves.

When an extremely young songbird, which is usually quite

naked or only sparsely feathered, is brought to us, we immediately put it into a box. A shoe box, or any container deep enough to keep it from falling out or scrambling out later, will do. We line the box with a mat of grass, excelsior, or torn-up bits of paper. This substituted nest material helps to keep the young bird clean, can easily be replaced when it is soiled, and gives the bird something to grasp with its feet. The material also helps it to keep its balance when it raises its head high and opens its mouth to be fed.

## KEEPING THE YOUNG BIRD WARM

For the first few days after they are hatched, young songbirds are "cold-blooded." When you touch them they will feel cool because their bodies are adjusted to the temperature of the surrounding air. When they are five or six days old, their body temperatures go up to what is near normal for an adult bird. The youngster is then "warm-blooded" and will feel warm to your touch. It is important, until your charge is fully feathered, to keep its nest box covered with a cloth, especially at night, and *always* to keep the bird out of a draft.

WARNING

Keep a young songbird's nest box covered day and night until the bird is fully feathered, and be sure the nest box is always sheltered from drafts.

The parents instinctively know this. For example, the female robin, during the first few days after her youngsters are hatched, will "brood" them frequently—settle down in the nest with her feathers spread over the young birds to keep them warm and dry. This is the way she guards them at night, and if the weather remains cool and damp, she may spend time brooding them every day and up until the moment when they leave the nest.

I think one of the most touching and yet most amusing sights in all nature is a parent bird trying to cover a nest filled with young ones so large that when they shift about under her they lift the brooding mother into the air and almost off the nest.

WARNING

Captive birds can become weak and ill from eating house or bluebottle flies.

RULE OF THUMB

It's a good idea to supplement a young bird's egg-bread-milk-cod-liver-oil diet with chopped green leaves, watercress and nastutium stems, and also a little cottage cheese now and then.

## WHAT TO FEED YOUNG SONGBIRDS

Our basic food for all songbird nestlings, other than hummingbirds, is equal parts of the finely mashed yolks of hard-boiled eggs (the yolks are more nourishing and easier for the birds to digest) and finely sifted bread crumbs, *slightly* moistened with milk or cod-liver oil. This mixture will agree with starlings, blue jays, cardinals, towhees, robins, catbirds, orioles, sparrows, blackbirds, waxwings, bluebirds, thrushes, and other small birds. Good supplementary foods are canned dog food, bits of grapes, cherries, raisins, bananas, soft apple pulp, pieces of earthworms that have been "squeezed out," and bits of scraped or finely chopped beef.

Many years ago I knew of a woman in Amsterdam, New York, who raised one of two young flickers entirely on bananas! The other flicker, besides eating bananas, ate boiled veal, strawberries, cherries, and boiled green peas. She also raised a young bank swallow on bread, milk, and hard-boiled egg yolk. When the swallow got older and refused these foods, she hired children to catch flies for it. In one afternoon the bird ate eighty-five large flies, but in our experience flies are not good for captive birds, which often become weak and ill after eating them. Possibly the house and bluebottle flies have too many germs on them.

It will greatly strengthen the young bird to supplement its egg-bread-milk-cod-liver-oil diet with the chopped green leaves and stems of watercress and nasturtiums, which are rich in calcium and vitamins. Feed it also a little cottage cheese now and then for added protein. Young woodpeckers (flickers, downy woodpeckers, etc.) should be fed a mixture of canned dog food and the basic, finely mashed egg yolks. Young hummingbirds get their protein from flying insects, but you can supplement their diet with a mixture of 28 ounces of white cane sugar in one gallon of water, as recommended by Paul W. Colburn, director of the Tucker Bird Sanctuary in Orange, California. You can add a drop of red food coloring to make the mixture more attractive to the bird. After about ten days give it its first protein by adding some finely sifted dried dog food to the sugar water. (See page 223 for another successful formula.)

## What to Feed Hawks and Owls

Hawks and owls, both young and adults, require raw meat, preferably with the fur or feathers on it, for these aid their digestion. Feed them on freshly caught mice, or poultry, and raw beef sprinkled with cod-liver oil, with which chicken feathers may be mixed.

To get a newly acquired hawk or owl to start eating, cut small strips of raw beef, a few inches long, and lay them across the bird's feet. This seems to attract its attention and it will usually reach down and pick up the meat. It may drop it and you may have to lay the meat across the bird's feet several times before it will bolt it down. Once it starts feeding it will usually continue until its crop or craw in the lower part of its neck is bulging with food. After its crop is filled, it will not accept more until the food has digested and passed into its stomach. Feed the young hawk or owl about three times a day, adults once a day.

**TRY THIS**

**If you have trouble getting a young hawk or owl to start eating, try laying strips of raw beef across the bird's feet.**

## How to Feed the Young Bird

A baby bird, with its eyes not yet open, will usually raise its head and automatically open its mouth for food whenever it hears you come near its box. Put the chopped egg yolk and bread crumbs in a small dish and use a narrow wooden spoon or, better, a small paint brush to pick up food on the tips of the bristles. Poke the food well back in the bird's throat but don't give it too much at one time, or it may choke. Continue feeding until it will not accept any more food. It should be well fed, especially at nightfall, just before it goes to sleep.

When a bird has had enough to eat, it will usually refuse to swallow. I have often watched parent robins feeding their young in the nest. They do not rotate their feeding among the birds, but push food down the throat of the one with its neck stretched highest, and its mouth opened widest. As soon as this youngster has had enough to eat, it will refuse to swallow. If the parent robin puts a worm or bug down the throat of the young one several times and it does not swallow, the parent will give it to the next robin in the nest, and so on, until all are fed. The rule to follow in feeding is: Don't put more food down a bird's throat until it has swallowed the food you have just given it.

RULE OF THUMB

**Wait for a bird to swallow any food you have just given it before putting more down its throat.**

Handle a young bird just as little as possible because merely holding it in your hands may sap its strength and kill it. Feed the young bird only during the daytime, starting early in the morning and ending at dusk, when you should cover the bird's box or cage to encourage it to go to sleep. The feedings should be frequent and *regular*, at least every fifteen to twenty minutes apart, or, at most, half an hour apart.

## How Much Will a Young Bird Eat?

Many years ago a bird scientist wanted to measure how much a young robin would eat at the time that it was ready to leave the nest. A young bird's appetite is probably greater then than at any other time of its life. The young robin of his experiment ate fourteen feet of earthworms on that last day of its nest life!

Another scientist discovered that parent robins will bring a total of about three pounds of food to their average brood of four young ones during the two weeks that they stay in the nest.

People who don't realize what enormous appetites young birds have may unknowingly allow the ones they adopt to starve to death. Young birds don't eat much at one time, but their food digests so rapidly that their parents feed them almost continuously from dawn until dark.

# Water and Sunshine

*Don't force a bird to drink!* When I was a boy, the first bird that I ever tried to raise, a young blue jay, died because I opened its mouth and poured water down its throat. Apparently the water got into its lungs, for it died within twenty-four hours. Small birds can be killed very quickly by forcibly giving them water. While they are in the nest, most young songbirds get sufficient water for their needs from the insects and wild fruits that the parent birds bring to them. In the same way, your young bird will get moisture from the food that you give it. If you occasionally feed it blackberries, mulberries, strawberries, and other small juicy fruits you will make certain that it gets enough liquid in its diet.

When the bird is old enough to sit on a perch, we offer it water in a shallow dish, from which it soon learns to drink. You may teach it to drink by dipping its bill in a cup of water and then holding the container of water in front of it, or by putting it somewhere within the bird's reach.

Give your young bird sunshine, but not too much of it. It is best to keep it shaded from the midday sun, which is usually too hot for a young bird and might kill it.

WARNING

Never force a bird to drink.

TRY THIS

A good way to teach a bird to drink is to dip its bill in a shallow bowl of water.

# How to Care for Older Birds

When we first get an ill or injured adult songbird or a young bird that is so well grown that it has already learned to fear man, we put it in a large cage or, preferably, in a room where it is free to fly. We give the grown-up the same kinds of foods that we give to a young bird because this will sustain it until it is strong enough to take care of itself. When we are sure that the bird can fly, we

WARNING

Always keep a young bird shaded from the strong midday sun, which is very hot and might kill it.

release it in our garden. There, unless it flies away immediately, we can watch over it.

RULE OF THUMB

**Feed adult birds three or four times a day.**

When they are first rescued, these older birds may not eat, even if we put food before them, because a bird that is already used to eating insects may not recognize our mixed egg yolk and bread as food. This means that we shall have to force-feed it, perhaps for a few days, until it will eat the food by itself. Some adult birds that we have had never voluntarily took food as long as we had them. Others learned to eat our offerings very quickly. We force-feed these adults about three or four times a day because a full-grown bird doesn't require nearly as much food as a nestling, which may eat its own weight in food each day. For the well-grown young bird or the adult, we put a shallow pan of water on the floor of the room or of the cage in which the bird is kept, so that it can drink or bathe whenever it chooses.

## FORCE-FEEDING A BIRD

One of the first things to learn in force-feeding a bird is how to hold it without injuring it. This is really easy. To learn the proper way to hold a bird, all you need to do is to pick up a large apple or an orange as if you were going to eat it.

Now look at the position of your fingers. They encircle the apple just in the way they should encircle the body and wings. Pick up the bird so that its back is against the palm of your hand and your fingers curve around its breast. The back of its head now lies comfortably in the V between your thumb and index finger. As you hold the bird, not tightly, but loosely in this position, you are holding its wings against its sides. It can't struggle, and it will automatically reach its feet out to grasp your little finger. It will now lie quietly and comfortably in your hand and you are ready to force-feed it.

Using the thumb and index finger of the hand that is holding the bird, gently squeeze them together against each side of its bill, at the *base* of the bill, and the bird will open its mouth. When it does, you have your free hand to feed it. Use a narrow wooden or plastic spoon. Make sure that the food is placed *in back of its tongue and well down its throat*, or it may spit the food out.

Withdraw the narrow spoon from the bird's mouth and it will almost immediately swallow, unless it is badly frightened. If it doesn't swallow, hold it quietly until it does. Continue feeding the bird until it altogether refuses to swallow. Patience is a virtue in force-feeding a bird. You must remember that this bright-eyed creature, whose wildly beating heart you feel pulsating in the palm of your hand, may be too terrified to swallow. *Be gentle, be patient, and be kind.* You will be surprised at how quickly the bird senses your good intentions and will respond to them.

**GREAT CRESTED FLYCATCHER**

| | |
|---|---|
| Number of Eggs Laid in Clutch: | 4-8 |
| Days to Hatch: | 13-15 |
| Days Young in Nest: | 15-18 |
| Number of Broods Each Year: | 1, possibly 2 |
| Lifespan: | 6 to 11 years |

*Great crested flycatchers range throughout the eastern United States, west to the prairies, and into southern Canada. They build cuplike nests of leaves and mosses in the natural cavities of trees, in abandoned woodpecker holes, and in birdhouses.*

# *More Tips from a Career Spent Attracting Birds*

For several good reasons, we remember the first winter we fed birds in the garden of our home in suburban New York City. Squirrels gave us our first "problem"—the big, sleek gray squirrels that practically overran the suburbs and public parks of New York, Boston, Philadelphia, and other large cities in the eastern United States, and had little natural habitat from which to feed.

In the country, squirrels had seldom visited my feeding trays because the nearest woodland was almost a quarter of a mile away. As any squirrel would tell you—if he could make himself understood—it is risky for him to travel far without the protection of a tree up which he can run to escape from hawks, owls, dogs, cats, and those men and boys who take up a rifle or shotgun to hunt squirrels in the fall of the year.

In the tree- and shrub-planted parks and suburbs of a large city, some of these hazards for squirrels don't exist, or they are reduced to a minimum. It is against the law to hunt within city limits, and the large hawks and owls that ordinarily feed on squirrels and help to keep down their population seldom live within or close to large cities. Without these checks on their numbers, gray squirrels have multiplied and become so fearless that they run about in our suburban neighborhood like domesticated animals.

We like to see squirrels in our yard. During the winter we feed them ears of corn and peanuts, which squirrels are so fond of. Squirrels are beautiful, graceful animals, and they are, to us, just as interesting to watch as are the birds. It was in the beginning, when the squirrels kept birds away from the feeders, that they became a problem.

After we had put up our feeding stations, gray squirrels got into them easily by climbing the wooden posts that supported them and by running down the wire with which we suspended one feeder from the limb of a tree in our yard.

When they got in the bird feeders, they sometimes sat there eat-

**COMMON GROUND DOVE**

| | |
|---|---|
| Number of Eggs Laid in Clutch: | 2-3 |
| Days to Hatch: | 12-14 |
| Number of Broods Each Year: | 1 |
| Lifespan: | 6 years |

*Common ground doves range throughout the southern and southwestern United States. They nest in depressions on the ground in fields, woods, or in a few twigs in a low bush or tree, 1 to 21 feet above the ground.*

ing for an hour or more. We didn't much mind how much they ate, but meanwhile songbirds, which are afraid of squirrels and won't come to the feeders while they are in them, stayed away. We wanted the squirrels in our yard, but we also wanted the birds. What could we do about it?

## *Metal Bands and Metal Cones to "Squirrel-Proof" the Feeders*

Now you can buy many "squirrel-proof" feeders, but not so back then. First, we nailed a strip of thin sheet metal completely around each of the wooden posts that supported our feeders. When the squirrels tried to get to them by running up the posts, they couldn't climb over the smooth metal surfaces. That stopped them only temporarily, for the squirrels soon showed us that they had other ways of reaching the feeders.

RULE OF THUMB

**To keep your feeders safe from squirrels, set them at least 60 inches above the ground and at least 8 feet away from any object from which a squirrel could jump.**

Squirrels will climb the sides of frame buildings and fences or climb out on the overhanging branches of trees and shrubs to leap down on the feeder from above. We soon found that we had to set every one of our feeding-station posts at least 8 to 10 feet away from any object that a squirrel could climb upon to get at a point above, or nearly above, the feeder. In the course of trying to protect our feeders from squirrels we also discovered that it was best to set feeders on posts at least 54 inches or, preferably, 60 inches above the ground. This kept them out of the reach of some of our more agile squirrels, which seemed to have springs in their feet.

To prevent squirrels from climbing down the wire to reach the feeder that we suspended from a tree limb, we had a tinsmith make a circular metal cone, about 30 inches in diameter. We slipped this down on the wire to a point about 18 inches above the feeder. The squirrels that ventured down the wire to the metal cone suddenly found themselves sliding down its steep sides until they pitched off it to the ground. After trying a few times and failing to reach the

### How to Keep Squirrels from Gnawing Bird Nesting Boxes

Gray squirrels like to get inside some of the bird nesting boxes, just as the birds do. There they will build a warm, leafy nest to sleep in, or perhaps to raise a litter of young ones in. In much the same way, squirrels in the forest enlarge the entrance holes to woodpecker nesting cavities in the dead stubs of trees.

Suburban squirrels, in gaining entrance to man-made bird boxes, follow an old, established squirrel custom. If your bird boxes are on trees where squirrels can gnaw them, this is easily prevented by nailing a metal ring, or a piece of sheet metal an inch or two wide, around the outside of the entrance hole to the bird nesting box. The strip also may be run into the bird box and nailed on the inside of the front panel so that the entrance hole inside and out is covered with the metal. This will discourage squirrels from gnawing at the edges of the hole to enlarge it. Make certain that the metal is smooth and that it has no sharp projecting edges, which can injure a bird going in and out of the nest box.

TRY THIS

You can safeguard a hanging feeder from squirrels by equipping the suspension wire with a circular metal cone about 30 inches in diameter.

feeder, they gave up and returned to those big yellow ears of corn that we heaped in their wire basket feeder that hung on the trunk of our white oak tree.

## PROTECTING FEEDING STICKS FROM SQUIRRELS

The squirrels also gave us a problem with our feeding sticks, which we had filled with a peanut butter mixture to attract chickadees and nuthatches. When the squirrels discovered them, they climbed down the wires to the feeding sticks to see what they were, and ate the peanut butter mix. We didn't mind them eating it, but they also gnawed away pieces of wood from around the edges of the feeding stick holes, which had become saturated with the peanut oil. If the squirrels continued to do this, they would destroy the feeding sticks

WARNING

If allowed, squirrels will gnaw at feeding sticks and destroy them in short order.

## Keeping Squirrels off the Window Shelf Feeder

Our first window shelf feeder was a plain board or tray that I fastened to the sill of one of our first-floor windows. The sill was only about 3 feet above the ground and squirrels easily leaped up or ran up the side of the house. To outwit them, we bought some heavy "fox wire" screening, because poultry wire was too pliable, and squirrels could bend and enlarge its openings. The fox wire had mesh openings about 1½ inches high and about
2 inches long. We enclosed our window shelf feeder with a piece of this, and bent it to form a rectangle about 18 inches high and 12 inches square at its ends. We tacked the bottom of this wire frame to the edges of the feeding shelf, and nailed a narrow strip of wood over the edges of the wire where they met the sides of the house. Chickadees, nuthatches, juncos, purple finches, goldfinches, tree sparrows, white-throated sparrows, and song sparrows passed easily through the openings of the wire mesh, but it kept out the larger squirrels, pigeons, starlings, blue jays, and grackles. Our chickadees flew directly through the openings, without touching the wire, but most birds alighted on the horizontal strands when coming into or leaving the shelf feeder.

If you don't want to build your window shelf feeder, some of the dealers in bird-attracting supplies sell glass-enclosed ones. We used one that had a long, narrow opening near the outside bottom of the feeder, which kept out squirrels and pigeons but allowed smaller birds to enter.

in a short time. We had to do something to prevent it.

Instead of having metal cones made to put on the wires above the feeding sticks, we simply stopped filling the holes in them with the peanut butter mixture and substituted suet. We had noticed that our squirrels had never eaten the suet that we occasionally tied to the limbs of trees for the birds. The peanut butter mix, which our birds liked so well, we put in small open compartments in those feeders that the squirrels couldn't reach. From that time on, after they discovered we had put suet in the feeding sticks, the squirrels no longer bothered them.

## How the Blue Jay Serves Other Birds

One fall day years ago my wife and I went out on our terrace in the backyard and sat down to watch our birds. I had filled the feeders with our wild-bird seed mixture and had scattered some on the ground under the shrubs along our north property-line fence for the flock of white-throated sparrows that had come to spend the winter with us. The white-throats liked to scratch about among the fallen leaves under the shrubs, and they seemed to prefer eating grain there to eating it in the feeders.

We sat watching half a dozen of them hopping about, picking up the seeds and cracking them in their bills, only pausing now and then to look about them and call *s-s-s-s-s-s-s-t-t-t!* in that lisping way in which they seem to talk among themselves. Two blue jays sat on the edge of our open feeder. This was fastened on top of a post set in the lawn about 15 feet from where the white-throats were feeding. The jays were eating sunflower seeds, and bolting them down one after another. The yard seemed peaceful.

Suddenly one of the jays straightened and raised its feathered crest. It looked toward the shrubbery under which the white-throats were feeding, then flew quietly toward them. It alighted in the top of the tallest shrub above the white-throats, looked down toward the ground, and screamed *y-a-a-a-n-n-h-h! y-a-a-a-n-n-h-h! y-a-a-a-n-n-h-h!*

The white-throats flew up into the branches of our oak tree and began calling excitedly. I got up and walked quietly toward the place where they had been feeding. A black cat jumped out from under the down-sweeping branches of one of our pine trees and ran out of our yard. It had been stalking the sparrows, and had been almost upon them when the blue jay's warning cry had sent them

| BLUE JAY | |
|---|---|
| Number of Eggs Laid in Clutch: | 3-6 |
| Days to Hatch: | 16-18 |
| Days Young Are in Nest: | 17-21 |
| Number of Broods Each Year: | 1 |
| Lifespan: | 5 to 15 years |

*The blue jay ranges east of the Rockies, from southern Canada to the Gulf States. Its bulky nest is made of twigs, sticks, bark, moss, leaves, and grass and can be found from 5 to 50 feet from the ground in vines, bushes, the crotch of a tree, or in branches close to the trunk.*

**AMERICAN ROBIN**

| Number of Eggs Laid in Clutch: | 3-6 |
|---|---|
| Days to Hatch: | 12-14 |
| Days Young in Nest: | 14-16 |
| Number of Broods Each Year: | 2 or 3 |
| Lifespan: | 4 to 12 years |

*The American robin ranges over North America from wilderness Canada and Alaska south to the Gulf coast and southern California. It builds its mud-lined nest of weeds and grasses between the ground and the treetops, but usually 5 to 20 feet up in the crotch of an elm or maple, on rail fences, posts, or porch gables, often on a garden arbor or on a shelf built for it.*

flying to safety. These little dramas of life or death for the birds were played in our yard every day. Invariably it was the blue jay that first warned the other birds of danger—of a cat or a threatening hawk or a shrike—even before the sharp-eyed, wary starlings discovered it. We called our blue jays our "watchdogs for the birds," for we knew that they have saved the lives of many of them with their loud, timely warnings.

## Housecleaning Your Bird Nesting Boxes

Each of our birdhouses has either one of its sides or the top of the box hinged. This enables us to open them and clean out the old nesting material at least once a year. If we didn't, the birds might not use them the following year. Besides, other animals smaller than birds might get in the nesting boxes in the fall or win-

### Birds That Fight Their Reflected Images

Almost every spring, one of the robins that nested in or near our yard discovered his image in one of our basement windows and started to fight it. Robins are strongly territorial, and each male will chase away other male robins that come within the invisible boundaries of the part of our yard that he has declared to be his own. If we didn't try to stop our robin, he would have jumped at his reflection in the window, sparring with it like a bantam rooster for a week or more. A robin nesting in the yard of one of our neighbors fought all one day with his image, which he saw reflected from the shining metal disk on one of the rear wheels of their car. To prevent the bird from exhausting himself, they kept their car in the garage the next day.

We stopped our robin from its wearing battle with our basement window by putting a fine-mesh window screen in front of it. Apparently this broke the image, for the bird stopped fighting its imaginary rival.

ter, before the birds return to nest in the spring. During several winters we have had some of the big-eyed white-footed mice that usually live in a nearby woodland move into one of our bird boxes. These are gentle little wild folk that remind us of the pixies of the fairy tales we read when we were children. They are harmless to birds, but a pair of wrens or bluebirds may not nest in the box if they find it already occupied by mice.

Once, when I lived in the country, a pair of flying squirrels made a nest in one of our flicker boxes. These small squirrels are so beautiful and were so interesting to us that we did not disturb them or the nest box until we found that they had deserted it. Then we cleaned it out thoroughly to prepare for the next family of flickers that later moved in.

## THE BLOWFLY—ENEMY OF NESTLING BIRDS

When we lived in New York, we always cleaned out our bird boxes about March 15, just before the songbird nesting season began. If we found a family of white-footed mice in one of them, we didn't feel so badly about evicting them at that time because winter was usually past. Some people clean out their birdhouses in the fall, after the nesting season is over, but we delayed ours until spring for a most important reason.

WARNING

An infestation of parasitic blowfly grubs can endanger your young hole-nesting birds.

The young of many hole-nesting birds are parasitized by an insect that in its young, or "grub," form may either directly or indirectly kill them. It is to help control this insect that we cleaned our birdhouses in the spring. Here's why.

Shortly after the hatching of young bluebirds, tree swallows, chickadees, crested flycatchers, and other birds that will nest in bird boxes, a blowfly, or bluebottle fly, may call on the young birds. The fly will lay its eggs on them and the tiny grubs, which hatch from the fly's eggs, attach themselves to the young birds and begin sucking their blood. In the summer of 1951 three young house wrens in one of our garden broods died from the attacks of these grubs.

### How to Build a Brush Pile

Some of our friends have been surprised to learn that there is a right and a wrong way to build a brush pile for birds. We would prepare for ours in advance by saving the limbs of trees and the branches of shrubs that we pruned in the spring and fall. These we piled in a far corner of the yard, under a leafy tree or shrub where they couldn't be seen during summer. Some of the heavier limbs were from trees that we might not cut once in four or five years. It was these that we used for the sturdy "foundation" of the brush pile, and they usually lasted for several years.

About November 1 we built the brush pile on the bare ground of our vegetable garden. We put down the heavy tree branches first, crisscrossing them until we had a tangled pile about 6 feet square and about 30 to 40 inches high. On top of these and in among the heavy branches we put lighter limbs until we had built the pile to about 5 feet above the ground. To finish it, or "top it off," we laid on small branches of evergreen trees—hemlock, spruce, balsam fir, and yew. If the weather was mild, we didn't put this final layer or "roof" on until after Christmas, when our discarded Christmas tree and those of our neighbors gave us plenty of material. These evergreen boughs protected the inside of the brush pile from snow and rain and made it a warm, dry, safe retreat that the birds spent much time in during cold weather. We often spread feed around the base of the brush pile for them when there was snow on the ground.

About April 1, or at any time before birds started nesting, we dismantled our brush pile. We burned the lighter branches, but saved the heavy ones for the new pile that we would build the following fall.

Fortunately for the young birds the grubs have an insect parasite that destroys them and thus helps the birds. It is a tiny chalcid fly (pronounced KAL-sid) that lays its eggs on the blowfly grubs. When the young of the chalcid fly hatch, they eat the grub that is sucking blood from the birds. Many of these chalcid flies sleep through the winter in the bird nest material within the bird box. If you destroy this material in the fall, especially if you burn it, you will destroy the valuable chalcid flies. That is why we always waited until early spring to dump the nest material out on the ground in some inconspicuous place where the chalcid flies could emerge to go on with their good work. Remember—do *not* burn the old nest material. Simply sweep it out on the ground just before the nesting season begins.

A bird scientist in New England discovered an even more effective way to get rid of the blowflies and, yet, save the chalcid flies, both of which may spend the winter in the same bird nesting material. After each brood of young birds has flown, he emptied the nest material out of the bird boxes into a metal container and covered this with a piece of fine-mesh window screening. When the blowflies hatched from the pupal cases in which they had transformed from grubs to adults, they were too large to escape through the fine-wire mesh, and so they died. When the smaller chalcid flies emerged, they could escape through the wire screening and were free to continue their work of parasitizing the blowfly grubs that attack young birds.

**SCARLET TANAGER**

| | |
|---|---|
| Number of Eggs Laid in Clutch: | 3-5 |
| Days to Hatch: | 13-14 |
| Days Young in Nest: | 9-11 |
| Number of Broods Each Year: | 1, possibly 2 |
| Lifespan: | 3 to 9 years |

*The scarlet tanager ranges throughout southern Canada and the eastern United States west to North Dakota and Oklahoma. It builds a shallow saucerlike nest of twigs, rootlets, and grasses out on the limb of an oak, ash, beech, or maple, 4 to 75 feet above the ground.*

## WASPS IN BIRDHOUSES

One summer day some years ago, a man in Nebraska who had a pair of wrens nesting in a bird box in his yard discovered one of them dead just inside the entrance hole. It was a female and, as he found out later, she had been stung to death by a wasp. Two days later the male got another mate and brought her to the nest box. She, too, was stung by a wasp as she entered the box and died just within the doorway. A colony of paper-making wasps had got established in the nest box, perhaps after the wrens had built their nest, and had killed two of the birds before the man discovered them. We had heard of many bird boxes in which wasps and birds had nested at the same time, but had never known a wasp to kill a bird until we learned of the Nebraska man's experience.

These wasps are called by scientists *Polistes*. They live in colonies and each colony builds a flat, open cluster of cells that the queen begins when she awakens from hibernation in the spring. These paper cells that make up the "nest" look like the comb of a honeybee and are the nurseries or chambers for the young wasps of the growing colony. The *Polistes* wasps build these nests suspended from the inside roofs of open sheds, inside bird boxes, and in other places that are sheltered.

The *Polistes* wasps, although able to sting painfully, are not as aggressive and as fiery as the white-faced hornets that nest in big gray paper nests. These nests are shaped like footballs, and the hornets build them suspended from the branches of trees and shrubs. The *Polistes* will allow you to approach them in their open nests quite close, if you move quietly and don't make sudden motions that anger them. Usually these wasps do not attack birds. In our few experiences with them in our bird nesting boxes, they got into them early in spring, before nesting had started, which made it much easier to get rid of them.

In the beginning, we used an aerosol bomb to destroy the wasps. After dark, when they were gathered quietly on the nest, we plugged the entrance hole with a cloth to prevent the wasps from getting out of the bird box. Then we shot a spray of an insecticide from the bomb into the box for about ten to fifteen seconds. The next morning when we opened the birdhouse, we found that every wasp within had been killed by the deadly liquid. Although we never found a dead bird in the nest box later, when it was occupied by a pair of wrens, I was nevertheless uneasy about the possible lasting lethal effects on them or their nestlings.

One August day I wrote to the U.S. Fish and Wildlife Service, Washington, D.C., and asked if there were any other means of evicting wasps from birdhouses without using an insecticide such as DDT, which was common at the time. The service wrote that they had made a series of experiments on the effects of spraying DDT inside the nest boxes of birds, or directly on the nests, eggs, or young. They found that DDT had no observable effect on the adult birds, but that it might have some effect on the survival of the nestlings. Later, of course, studies showed DDT's terrible effects on bird populations when it entered the food chain. Rachel Carson's *Silent Spring* tells this story. Instead of using DDT, the service advised Sevin, a carbamate with which they had been experimenting. It gave excellent control of insects within the nest boxes, and was relatively nontoxic to birds; however, it can be extremely toxic to bees. They suggested that, used within bird boxes, the compound would not threaten bees, which do not live inside of bird nesting boxes as wasps do, and would be useful for insect control within the birdhouses early in the year, before the birds' nesting season has begun. The service does not advise using Sevin during the time that the young are in the nest. The boxes become warm inside during the day and the insecticide might subject the nestlings to some respiratory intoxication; however, it had not been observed in their studies.

In evicting wasps from bluebird houses along my bluebird trail in North Carolina, I did not use an insecticide, but opened the front of the bird box and then assaulted the wasps and their nest with a blunt stick. However, in this method there was always a risk of getting stung. After the nest was destroyed, the wasps would remain out of the nest box temporarily but would try to renest in it again. They had to be evicted repeatedly until a pair of birds was in possession, when there was less likelihood of the wasps nesting in the box.

## A Flicker Problem

Although we never experienced it, friends of ours who live in the country once had a flicker come to their house on a fall day and start drilling a hole in one of its outside walls. What was the bird's purpose, they asked, and what could they do to stop it?

**HAIRY WOODPECKER**

| | |
|---|---|
| Number of Eggs Laid in Clutch: | 3-6 |
| Days to Hatch: | 14 |
| Days Young in Nest: | 28-30 |
| Number of Broods Each Year: | 1 |
| Lifespan: | 7 to 13 years |

*The hairy woodpecker nests in all types of forests, from the tree limit in Alaska and Canada south to Florida and the Gulf coast. It excavates a cavity mostly in living maple, oak, apple, aspen, and beech trees, but sometimes in the tops of dead trees. The hairy woodpecker hole has an entrance 1 1/2 to 2 inches in diameter and is located 5 to 30 feet or more up in the tree with a soft bed of fresh wood chips in the bottom. It will also nest in a birdhouse designed for it.*

In autumn, flickers and other woodpeckers will drill holes in the dead stubs of trees, where they hollow out a place to sleep during the cold winter nights. Once in a while, if there are no dead stubs of trees about, they will drill into the sides of barns and other buildings to gain a sleeping place for the winter.

As with the flicker that had drilled holes in the barn, to solve the problem, we advised our friends to put up at least one flicker nesting box in a tree that stood near their house. Then, if it didn't abandon its hole-making for the flicker box, they should frighten it away by firing a gun into the air, which they eventually tried. Although they had to fire the gun several times before the flicker would abandon the idea of roosting in their attic for the winter, the bird finally left. Obviously, a gun wouldn't have been an option in the city or suburbs.

There is a happy sequel to the story. The following spring, a pair of these birds nested in the box our friends had put up, and during the following winter a flicker roosted in the box each night. Whether one of these was the same flicker that had tried so hard to break into their house they never knew. But they have a much kindlier feeling toward flickers now that they understand them and have them for their nearest neighbors.

## Vines to Cover Your Fireplace Chimney

One of our neighbors liked birds and was greatly interested in them. Yet he admitted to me one day that he had made a mistake when he planted an English ivy vine at the base of his fireplace chimney. After twenty-five years, the vine had covered the outside brickwork from ground level to chimney top with a luxuriant growth of dark-green leaves. The vine was beautiful and added greatly to the attractiveness of his home, but he said, "It had become too noisy with roosting birds."

House sparrows and starlings swarmed into the vine to sleep, especially during the winter, because of the warmth of the chimney, and the evergreen leaves that protected them from wind, rain, and snow. Our neighbor, who was a light sleeper, said that early in the morning the chatter of birds, which liked to linger on cold days in their warm shelter against the chimney, kept him awake.

No matter how much you may be interested in birds, it would be well for you to consider our neighbor's problem. If you, too, are a light sleeper, it might not be wise to risk losing your ardor for the birds by inviting them to sleep just outside your window. We had an ivy vine on the outside of our chimney, and we didn't mind the early morning chatter of the birds because we, too, were usually early risers. If you like to sleep late, you might plant your vine at the base of a large tree or a tree stump, somewhere in your yard away from the house. There the birds will have their roost and you will get your proper rest—an arrangement that our neighbor said, rather ruefully, he wished he had made.

## Cats and Birds

I don't know of any problem in bird-attracting that is more perplexing to some people than the one of cats and birds. I have known a number of families who hesitated about starting to attract birds because they owned a cat. They were in honest doubt about the wisdom of drawing birds to their yard while they had a pet whose natural instincts are to hunt and kill. I know people who have cats and stoutly maintain that they never kill birds; others who have them are equally sure that they do; and some of the people who don't own cats and attract birds are positive that cats kill little else but birds.

Our position on this vexing question is, we hope, a sensible one. We are very fond of cats, and have had several of them during the years we have been attracting birds. We don't believe that cats, in

### Grosbeak Usurpers

At times, evening grosbeaks and other northern finches may swarm over the feeders in the garden, and the evening grosbeaks, especially, may usurp them and temporarily exclude from them some of the smaller garden birds. The grosbeaks, however, like to feed high above the ground, and you may lure them away from the main feeders in the garden by attaching an open feeding tray to one of your second-story windowsills (see window shelf feeder, page 23). Keep this feeder well supplied with plenty of sunflower seeds, and you will have the pleasure of watching these handsome birds, closeby, through your window and of knowing that they are not hindering the feeding of the smaller chickadees, nuthatches, native sparrows, and other small birds at your garden feeders.

general, can ever be taught not to hunt birds, but if you keep your cat where it will not be able to reach them, your garden birds should be safe from it. We kept our cat within a screened porch most of the time during summer. When someone in the family took it out for exercise, they put it on a long leash. Although this prevented our cat from catching birds, it did not stop our neighbors' cats from occasionally catching them.

We don't own a cat now but, having had one, we appreciate the affection that people may have for them. A cat stalking one of your favorite chickadees or cardinals may arouse your fury, but you should remember that the cat may be someone's pet—perhaps a child's—and any harm you bring it may cause a neighborhood quarrel and possibly a lasting regret over any hasty action on your part.

Our neighbors cooperated wonderfully in helping us with the cat problem. They "belled" their cats—put a small bell on the cat's collar, which tinkled with its every movement when it tried to stalk birds. This helped, but they also told me to chase their cats and to frighten them off in any way that we thought would teach their pets to stay out of our yard. By repeatedly chasing them away, whenever we saw them, we did not have to use violence and the neighborhood cats learned to avoid our yard, at least during the day.

A "cat-proof" fence, 6 feet high and made of strong woven wire, with 1½-inch mesh openings, will keep cats, dogs, and almost all other creatures out of your yard. The fence should overhang 2 feet at its top *toward the outside*, to prevent any animal from climbing over it. These fences are expensive; but, if you can afford one, they will give your songbirds complete protection against all animals, except predatory birds—the small hawks, owls, and shrikes that sometimes kill and eat songbirds.

# The Birds That Kill to Live

My wife and I shall never forget one cold February day a few years ago when we counted 346 songbirds feeding in our backyard. Red-winged blackbirds, grackles, cowbirds, house sparrows, and starlings fairly swarmed over several of the open feeders; downy woodpeckers, nuthatches, and chickadees were eating suet from our feeding sticks; and a mourning dove, a flock of juncos, and white-throated, fox, tree, and song sparrows pecked at the ground under shrubs where we had spread seeds for them. Four blue jays flew in and away from the feeders, carrying peanuts up into our white oak tree, where they sat holding them under their feet and hammering them open with their beaks. The calling of the red-wings, cowbirds, grackles, and starlings made a bedlam of creaking, whistling bird sounds that our neighbors could hear a block away.

We were watching the birds out of the window when we heard a jay scream a warning. Instantly the birds in our yard fell silent and flattened themselves on the feeders, on the ground, against trees, or wherever they happened to be.

Then, with the suddenness of a stroke of lightning, a small brown hawk swooped down out of the air and darted at the birds in one of the feeders. Just before it reached them, songbirds exploded upward all over the yard with a roar of wings that must have been as startling to the hawk as it was to us. The hawk, disconcerted by the birds that were in the air all about him, seemed to bound straight up into the air, turned sideward, and shot down again toward the feeder. That moment of indecision by the hawk was all that the songbirds had needed. Not one remained in sight. All had dived into shelter—into spruce trees, yews, and pines.

The American kestrel—for that is what it was—alighted atop a brush pile that we had built in our garden about fifteen feet away from the most exposed open feeder. I looked at the bird through my binoculars and saw him in all his magnificent wildness, his reddish brown feathers puffed out, his tail pumping up and down, his black eyes blazing eagerly as he looked about. For several moments he sat there, then sprang upward into the air and with a piercing cry flew rapidly away.

I almost felt sorry for the hawk, for his seeming disappointment at missing a meal on that cold day, had I not remembered what great little mousers these American kestrels are. In the grassy fields of our suburban neighborhood the hawk would not need to hunt long in order to catch a field mouse—a happy substitute for one of our garden birds.

For two minutes after the hawk had flown away, not a bird came out of hiding, except our bold little chickadees, which seem to have little fear of hawks. I walked out to the brush pile I had built for the birds for just this kind of emergency. I kicked it lightly and 14 birds—juncos, white-throated sparrows, fox sparrows, and house sparrows—flew out. The brush pile had undoubtedly saved the life of at least one of them on that day, as it has on other days since.

Besides attracting small hawks in winter, the songbirds at your feeders will attract shrikes—another kind of predatory bird. Shrikes are usually gray, black, and white, about the size of a robin or a little smaller. Females and immatures are a dull *brown*, black, and white. Shrikes sometimes kill robins, mockingbirds, blue jays, and other medium-sized birds, but they more often take the smaller starlings, house sparrows, juncos, goldfinches, crossbills, siskins, bushtits, and others of the 4- to 8-inch sizes. All shrikes will catch and eat songbirds when they can't find mice or the grasshoppers and crickets that make up a large part of their food in summer.

Although small birds are sometimes killed by shrikes at the feeders, they can escape (1) by flying straight up into the air, keeping above the shrike; (2) by out-dodging it; (3) by darting into thick cover; (4) by remaining perfectly still when a shrike appears.

A shrike chased a brown creeper into the yard of friends of ours and the creeper escaped by alighting on the bark of a tree and remaining motionless. The shrike, which had been flying closely after it, flew up into the tree and looked down at the bark intently for fully five minutes. The motionless creeper blended so well with the bark of the tree that, apparently, the shrike didn't see it and finally flew away. A minute later, when the creeper was sure that the shrike had gone, it moved up the tree, spiraling its way up the trunk as if nothing had happened.

**AMERICAN KESTREL**

| | |
|---|---|
| Number of Eggs Laid in Clutch: | 4-5 |
| Days to Hatch: | 29-30 |
| Days Young in Nest: | 27-33 |
| Number of Broods Each Year: | Usually 1 |
| Lifespan: | 10 to 17 years |

*The American kestrel ranges throughout Canada, Alaska, and most of the United States. It prefers to nest in the old tree-nesting holes of flickers, but it will also nest in the hollows of cactus, niches in the walls of city buildings and cliffs, and in birdhouses.*

## WHAT SHOULD YOU DO ABOUT HAWKS AND SHRIKES IN YOUR GARDEN?

In all the years that we have been feeding birds, we have never had a hawk or shrike that we couldn't get rid of by frightening it off or by live-trapping, if it became too persistent. We have never killed a hawk or shrike that has been attracted to our garden by our birds. When one comes to our yard and stays, we stop putting grain in our feeders where the birds would be exposed to attacks by hawks or shrikes. We scatter the grain, instead, under the shrubs and evergreen trees where our songbirds can feed close to protective cover.

People who operate bird-banding stations for the U.S. Fish and Wildlife Service sometimes catch small hawks and shrikes in their bird-banding traps. These wire traps are usually set by banders in their backyards to catch songbirds, which they band and then release in order to trace their migrations and length of life. Hawks or shrikes sometimes enter these traps to catch the songbird that may be hopping about, unharmed, within it, eating the grain with which the trap is baited to lure the songbird inside. After the bird-bander removes the hawk or shrike from the trap, he usually bands it, carries it at least ten miles away, and releases it. Hawks or shrikes released that far away seldom return to the bird-bander's yard. Bird-banders in recent years have favored "mist nets," which they stretch across openings in woodlands or garden paths to entangle small birds. The banders work carefully to untangle the

**DOWNY WOODPECKER**

| | |
|---|---|
| Number of Eggs Laid in Clutch: | 4-5 |
| Days to Hatch: | 12 |
| Days Young in Nest: | 21-24 |
| Number of Broods Each Year: | 1 or 2 |
| Lifespan: | 4 to 10 years |

*The downy woodpecker ranges from southern Alaska, across Canada, in suitable places in most of the United States, south to the Gulf of Mexico and Florida. Its nests are usually excavated in dead stubs of trees with a round entrance hole 1 1/4 inches in diameter, 5 to 50 feet above the ground, generally in a dead snag, but sometimes in a solid branch. The downy woodpecker will also nest in a birdhouse designed for it.*

birds from the nets, and after examination to record scientific information, band and release them.

We believe that if we are to attract birds and remain on good terms with all nature and with all birds, we must not allow ourselves to become prejudiced against any of the wild creatures that come to our garden. We do everything possible to protect our birds against the hawks or shrikes that attempt to kill them. But in doing so we remember that we are responsible for attracting the large population of birds that has, in turn, attracted the hawks and shrikes.

When you feed birds in your garden, this phenomenon—that one bird, a hawk or a shrike, must kill another bird to keep life in its own body—may be the hardest for you to accept. If you don't accept it, your enjoyment of birds for what they are—free, wild, uninhibited creatures—will be less and your breadth of view narrowed.

I shall not ask you to harden your feelings against the death of birds in your garden, even if that were possible for you. I *shall* ask you to look on predatory birds as creatures that are living as nature has directed them to live, and to realize that the action of the hawk that strikes down a sparrow at your feeder is no different from that of the sparrow itself, which only a moment before killed a beetle to satisfy its own hunger.

To live is to die, and one creature in your garden lives only until that day that it must give up its life to another—for another. Like the tides that roll upon an ocean beach, the lives of birds in your backyard will ebb and flow. But life—a rich, fascinating, varied life—will always be with you while your trees, shrubs, food, and water are there to say *Come and live with us!* to the songbirds in your garden.

# Appendices

## SOME DEALERS IN BIRD-ATTRACTING SUPPLIES

*(Any of these companies would be happy to send you a catalog, and many sell birdseeds as well as bird-houses and feeders.)*

American Birding Association, Inc.,
2812 W. Colorado Avenue #103, Colorado Springs, CO 80904

Audubon Workshop,
1501 Paddock Drive,
Northbrook, IL 60062

Bird Seed Savings Day,
Nature Center Association Inc.,
16 Holmes Street,
Mystic, CT 06355

The Bird Tree,
5 Swallow Lane,
St. Paul, MN 55110

Crow's Nest Book Shop,
Cornell Laboratory of Ornithology,
159 Sapsucker Woods Road,
Ithaca, NY 14850

Dialbird,
554 Chestnut Street,
Westwood, NJ 07675

Droll Yankees, Inc.,
27 Mill Road,
Foster, RI 02825-1366

Duncraft,
33 Fisherville Road,
Penacook, NH 03303

Hallco Products,
718 First Avenue,
Grand Rapids, MN 55744

Hummingbird Heaven,
454 Carmel Drive,
Simi, CA 93065

Hyde Bird Feeder Company,
56 Felton Street,
P.O. Box 168,
Waltham, MA 02254

Massachusetts Audubon Society
(The Audubon Shop),
208 South Great Road,
Lincoln, MA 01773

Nature House Inc.,
Purple Martin Junction,
Griggsville, IL 62340

Nelson Mfg. Co.,
3049 12th Street,
Cedar Rapids, MN 55744

Postmart,
Box 473,
Avon, CT 06001

Welles L. Bishop Co.,
1245 E. Main Street,
Meriden, CT 06450

Wild Bird Supplies,
4815 Oak Street,
Crystal Lake, IL 60014

Wildlife Refuge,
8845 Bath Road,
Laingsburg, MI 48848

## WHAT TO FEED BIRDS

*Many of the foods that appeal to birds are listed below along with the birds that are most attracted to them.*

### Beef Suet

Blackbird, red-winged
Bluebird, eastern
Bushtit
Catbird, gray
Chickadee, black-capped
Chickadee, Carolina
Chickadee, chestnut-backed
Creeper, brown
Crossbill, white-winged
Flicker, yellow-shafted (common)
Goldfinch, American
Grackle
Grosbeak, rose-breasted
Jay, blue
Jay, gray
Juncos
Kinglet, golden-crowned
Kinglet, ruby-crowned
Mockingbird
Nutcracker, Clark's
Nuthatch, red-breasted
Nuthatch, white-breasted
Oriole, Baltimore (northern)
Ovenbird
Owl, screech
Robin, American
Sapsucker, yellow-bellied
Shrike, northern
Siskin, pine
Sparrow, fox
Sparrow, tree
Sparrow, white-throated
Starling
Thrasher, brown
Thrasher, curve-billed
Thrush, hermit
Thrush, Swainson's
Thrush, wood
Titmouse, tufted
Towhee, rufous-sided
Warbler, myrtle (yellow-rumped)
Warbler, orange-crowned
Warbler, pine
Woodpecker, downy
Woodpecker, hairy
Woodpecker, pileated
Woodpecker, red-bellied
Wren, Bewick's
Wren, cactus
Wren, Carolina
Wren, house
Wren, winter

(If you find beef suet difficult to obtain, use one of the peanut butter mixtures that don't contain suet recommended in chapter 1.)

### White Bread

Blackbird, red-winged
Bluebird
Bunting, painted
Cardinal
Catbird, gray
Chickadees
Cowbird, brown-headed
Creeper, brown
Finch, house
Grackle
Jays
Juncos
Longspur, Lapland
Mockingbird
Nuthatch, white-breasted
Pheasant
Quail
Redpolls
Robin, American
Shrikes
Sparrow, black-throated (desert)
Sparrow, chipping
Sparrow, field
Sparrow, fox
Sparrow, song
Sparrow, tree
Sparrow, white-crowned
Sparrow, white-throated
Starling
Tanager, scarlet
Tanager, summer
Thrasher, Bendire's
Thrasher, brown
Thrasher, curve-billed
Thrush, hermit
Thrush, Swainson's
Thrush, wood
Titmouse, tufted
Towhee, rufous-sided
Warbler, yellow-throated
Wren, Bewick's
Wren, cactus
Wren, house
Wren, winter

### Cornbread

Blackbird, red-winged
Bluebirds
Cardinal
Catbird, gray
Cowbird, brown-headed
Jay, blue
Juncos
Kinglet, ruby-crowned
Mockingbird
Nuthatch, white-breasted
Robin, American
Sparrow, chipping
Sparrow, field
Sparrow, white-crowned
Thrasher, brown
Titmouse, tufted
Woodpecker, downy
Wren, Bewick's
Wren, Carolina
Wren, house

### Sunflower Seeds

Blackbird, rusty
Brambling
Bunting, painted
Cardinal
Chickadees
Cowbird, brown-headed
Crossbill, red
Crossbill, white-winged
Dickcissel
Dove, mourning
Finch, house
Finch, purple
Goldfinches
Grackle
Grosbeak, blue
Grosbeak, evening
Grosbeak, rose-breasted
Jay, blue
Jay, scrub
Juncos
Nuthatch, red-breasted
Nuthatch, white-breasted
Quail, bobwhite
Quail, Gambel's
Redpolls
Siskin, pine
Sparrow, chipping
Sparrow, fox
Sparrow, tree
Sparrow, song
Sparrow, white-crowned
Sparrow, white-throated
Thrasher, brown
Titmouse, tufted
Towhee, rufous-sided
Waxwing, cedar
Woodpecker, downy
Woodpecker, hairy
Woodpecker, red-bellied
Woodpecker, red-headed
Wren, Carolina

### Millet Seeds

Blackbird, red-winged
Blackbird, rusty
Bunting, indigo
Bunting, painted
Dickcissel
Dove, common ground
Dove, mourning
Finch, house

Finch, purple
Goldfinches
Jay, blue
Juncos
Longspur, Lapland
Quail
Redpolls
Siskin, pine
Sparrow, chipping
Sparrow, field
Sparrow, fox
Sparrow, house
Sparrow, song
Sparrow, tree
Sparrow, vesper
Sparrow, white-crowned
Sparrow, white-throated
Thrasher, brown

**Whole Corn**

Cardinal
Chickadees
Dove, mourning
Dove, rock (domestic pigeon)
Grackle
Grosbeak, blue
Grouse, ruffed
Jay, blue
Nuthatch, white-breasted
Pheasant
Quail
Turkey, wild
Woodpecker, red-bellied

**Whole Oats**

Blackbird, red-winged
Blackbird, yellow-headed
Bunting, snow
Cardinal
Chickadees
Dickcissel
Dove, mourning
Dove, rock
Finch, purple
Flickers
Grouse, ruffed
Juncos
Meadowlarks
Quail
Sparrow, house
Towhees

**Wild-Bird Seed Mixture**

*(millets, rape, canary seeds, sunflower seeds, peanut hearts):*
Blackbird, red-winged
Bunting, snow
Cardinal
Cowbird, brown-headed
Finch, house
Finch, purple
Goldfinches
Grosbeak, pine
Juncos
Lark, horned
Longspur, Lapland
Redpolls
Siskin, pine
Sparrow, field
Sparrow, fox
Sparrow, house
Sparrow, song
Sparrow, tree
Sparrow, white-crowned
Sparrow, white-throated

**Suggested Birdseed Mixture of 75 Pounds**

Millet seeds
25 pounds
Sunflower seeds
25 pounds
Buckwheat
10 pounds
Peanut hearts
10 pounds
Grit *(coarse white sand or ground oyster shells)*
5 pounds

**Scratch Feed**

*(coarse for larger birds; fine for smaller ones):*
Blackbird, red-winged
Blackbird, rusty
Bunting, painted
Bunting, snow
Cowbird, brown-headed
Dickcissel
Dove, ground
Dove, mourning
Finch, house
Finch, purple
Goldfinches
Grackle
Grouse, ruffed
Jay, blue
Juncos
Lark, horned
Longspur, Lapland
Pheasant
Pyrrhuloxia
Quail, bobwhite
Quail, Gambel's
Siskin, pine
Sparrow, chipping
Sparrow, field
Sparrow, fox
Sparrow, house
Sparrow, song
Sparrow, tree
Sparrow, white-crowned
Sparrow, white-throated
Starling
Thrasher, Bendire's
Thrasher, brown
Turkey, wild
Woodpecker, downy
Woodpecker, pileated
Woodpecker, red-bellied
Woodpecker, red-headed

**Peanut Butter**

*(mixed with suet and corn-meal) or Marvel-Meal:*
Bushtit
Chat, yellow-breasted
Chickadee, black-capped
Chickadee, Carolina
Chickadee, chestnut-backed
Creeper, brown
Finch, house
Flickers
Goldfinches
Jay, blue
Jay, Steller's
Juncos
Kinglets
Nuthatches
Redpolls
Robin, American
Siskin, pine
Sparrow, fox
Sparrow, song
Sparrow, tree
Tanager, western
Thrasher, brown
Thrasher, curve-billed
Thrush, hermit
Thrush, varied
Thrush, wood
Towhee, brown
Warbler, Audubon (yellow-rumped)
Warbler, orange-crowned
Warbler, Tennessee
Woodpecker, downy
Woodpecker, hairy
Wren, Carolina

**Walnut Meats**

*(black or English, crumbled):*
Cardinal
Catbird, gray
Chickadees
Finch, house
Goldfinches
Jays
Juncos
Nuthatch, red-breasted
Nuthatch, white-breasted
Siskin, pine
Sparrow, house
Sparrow, song
Sparrow, white-crowned
Thrasher, brown
Titmouse, tufted
Warbler, myrtle (yellow-rumped)
Woodpeckers
Wren, Carolina

### Doughnuts

*(suspended whole from tree branches or crumbled in feeders)*:
Bluebirds
Bushtit
Chat, yellow-breasted
Chickadee, black-capped
Chickadee, Carolina
Crossbills
Finch, house
Finch, purple
Flicker, red-shafted (common)
Jay, blue
Juncos
Kinglet, ruby-crowned
Mockingbird
Nuthatch, red-breasted
Nuthatch, white-breasted
Robin, American
Sapsucker, yellow-bellied
Sparrow, chipping
Sparrow, golden-crowned
Sparrow, song
Sparrow, white-crowned
Sparrow, white-throated
Tanager, western
Thrasher, brown
Thrush, hermit
Warbler, Audubon's (yellow-rumped)
Warbler, orange-crowned
Wren, Bewick's
Yellowthroat (a warbler)

### Fruits

**Sliced pieces of raw apple**
Chat, yellow-breasted
Finch, house
Jay, blue
Mockingbird
Orioles
Robin, American
Sparrow, white-crowned
Thrasher, curve-billed
Thrush, hermit
Warbler, orange-crowned
Waxwing, cedar
Woodpecker, hairy
Wren, cactus

**Baked apple**
Bluebirds
Robin, American

**Frozen crabapples**
Finch, purple
Robin, American
Waxwing, cedar

**Grapes**
*(either wild or cultivated; offer in sliced halves in feeders—white seedless are especially attractive to birds)*:
Bluebirds
Catbird, gray
Finch, house
Grosbeak, black-headed
Grosbeak, rose-breasted
Jay, gray
Mockingbird
Oriole, Baltimore (northern)
Oriole, hooded
Oriole, Scott's
Robin, American
Tanager, western
Thrasher, California
Thrush, Swainson's
Towhees
Warblers
Waxwing, cedar
Woodpecker, acorn

**Grape jelly**
*(may be put on feeding tray in special container to keep it from mixing with other bird foods; according to John V. Dennis, it does not freeze in winter)*:
Oriole, Baltimore (northern)
Sapsucker, yellow-bellied
Warbler, myrtle (yellow-rumped)

**Bananas**
*(peeled, then may be cut in pieces or put whole on feeding tray)*:
Bunting, indigo
Catbird, gray
Chat, yellow-breasted
Finch, house
Jay, gray
Mockingbird
Starling
Tanager, western
Warbler, Tennessee
Woodpecker, hairy
Wren, Carolina

**Strawberries**
Bluebirds
Catbird, gray
Grosbeak, black-headed
Grosbeak, rose-breasted
Mockingbird
Quail, bobwhite
Robin, American

**Bayberries**
*(cut fruiting branches from bushes, then put berries on feeding shelves or in tray feeders)*:
Chickadees
Warbler, myrtle (yellow-rumped)

**Oranges**
*(cut in halves and impale on sharp twigs of trees or on nails with points up, projecting upward from porch railings or from a special feeding board)*:
Catbird, gray
Grosbeak, rose-breasted
Mockingbird
Oriole, Baltimore (northern)
Tanager, scarlet
Tanager, summer
Tanager, western
Thrasher, brown
Warbler, myrtle (yellow-rumped)
Woodpecker, red-bellied

**Cherries**
*(canned or fresh)*:
Catbird, gray
Robin, American
Starling
Tanagers
Thrushes
Waxwing, cedar

**Cranberries**
*(fresh or canned)*:
Catbird, gray
Grosbeak, pine
Robin, American
Starling

**Watermelon**
*(pulp or rind)*:
Finch, house
Grosbeaks
Mockingbird
Orioles
Warblers

### Miscellaneous

**Baking powder biscuits**
*(short and crumbly)*:
Bluebirds
Robin, American

**Dog biscuit**
*(ground)*:
Bunting, snow
Chickadees
Grackle
Jay, blue
Junco, dark-eyed (slate-colored)
Lark, horned
Longspur, Lapland
Nuthatch, white-breasted
Sparrow, tree
Starling

**Cornbread**
Eaten dry by red crossbill but usually offered to birds in mixture with suet and peanut butter.

**Cake, cracker, and cookie crumbs**
Catbird, gray
Jay, blue
Jay, gray
Sparrow, chipping
Sparrow, house
Titmouse, tufted

**Piecrust**
*(dry)*:
Bluebirds
Catbird, gray
Chickadees
Grackle
Junco, dark-eyed (slate-colored)
Oriole, Baltimore (northern)
Starling
Tanager, scarlet
Tanager, summer
Titmouse, tufted
Wren, house

**Unbaked dough**
Jay, blue

**Cottage cheese, or pot cheese**
Bluebirds
Catbird, gray
Robin, American
Wren, Carolina

**Fried, mashed, or baked potatoes**
Catbird, gray
Crow, common
Grackle
Jay, blue
Starling
Thrasher, curve-billed
Wren, cactus

**Fresh tomatoes**
*(halved and put on open feeder)*:
Catbird, gray
Pyrrhuloxia
Sparrow, white-crowned
Warbler, Tennessee

**Eggshells of poultry**
*(finely crushed, may be put on bird feeder or pieces scattered on lawn)*:
Jay, blue
Martin, purple

**Barrel cactus seeds**
*(in Southwest)*:
Bunting, lark
Finch, house
Sparrow, black-throated (desert)
Sparrow, white-crowned
Towhee, green-tailed

**Cantaloupe seeds**
Nuthatch, red-breasted
Nuthatch, white-breasted
Sparrow, house

**Ground pumpkin seeds**
Chickadees
Junco, dark-eyed (slate-colored)
Nuthatch, white-breasted
Sparrow, tree
Warbler, myrtle (yellow-rumped)

**Broken squash seeds**
Chickadees
Junco, dark-eyed (Oregon)
Nuthatch, white-breasted

**Thistle seeds:**
Goldfinch
Siskin

**Pecan meats**
*(broken or ground)*:
Blackbird, red-winged
Bluebirds
Bunting, indigo
Catbird, gray
Chickadees
Cowbird, brown-headed
Dove, ground
Finch, purple
Goldfinches
Jays
Juncos
Kinglet, ruby-crowned
Robin, American
Siskin, pine
Sparrow, field
Sparrow, white-crowned
Thrush, hermit
Titmouse, tufted
Woodpeckers
Wren, Carolina
Wren, house

**Rice**
Dove, mourning
Quail, bobwhite
Sparrow, white-crowned

**Soybeans**
Dove, mourning
Grouse, ruffed
Meadowlark
Pheasant
Quail

**Rolled oats**
Junco, dark-eyed (Oregon)
Pigeon, band-tailed
Siskin, pine
Thrasher, curve-billed

**Chaff**
*(barn floor sweepings)*:
Bunting, snow
Lark, horned
Longspur, Lapland
Quail
Redpolls
Sparrow, savannah
Sparrow, tree
Sparrow, vesper

**Sugar water**
Catbird, gray
Hummingbirds
Mockingbird
Orioles
Thrasher, brown
Woodpecker, red-bellied

## SOURCES FOR MORE INFORMATION ABOUT ORNAMENTAL PLANTINGS FOR BIRDS

### U.S. Government Pamphlets

The following documents about plantings can be ordered from the U.S. Government Printing Office, Washington, D.C. 20420, and were available at the time of this edition of *Songbirds in Your Garden* (1994).

For information, or help locating a document, call the Office of the Superintendent of Documents, (202) 783-3238.

*Commerical Blueberry Growing*, U.S. Department of Agriculture, Farmer's Bulletin 2254. Cost: $1.50.

*Growing Blackberries*, U.S. Department of Agriculture, Farmer's Bulletin 2160. Cost: $1.25.

*Growing Raspberries*, U.S. Department of Agriculture, Farmer's Bulletin 2165. Cost: $0.90.

*Invite Birds to Your Home: Conservation Plantings for the Northwest*, U.S. Department of Agriculture, Soil Conservation Service. Cost: $1.50.

*Invite Birds to Your Home: Conservation Plantings for the Southeast*, U.S. Department of Agriculture, Soil Conservation Service. Cost: $1.00.

*Selecting Shrubs for Shady Areas*, U.S. Department of Agriculture, Agricultural Research Service, National Arboretum. Cost: $0.30.

*Shrubs, Vines, and Trees for Summer Color*, U.S. Department of Agriculture, Home and Garden Bulletin 181. Cost: $0.35.

*Transplanting Ornamental Trees and Shrubs*, U.S. Department of Agriculture, Home and Garden Bulletin 192. Cost: $0.70.

(Note: Similar pamphlets on conservation plantings for other sections of the country, not currently available from the Government Printing Office, may be available at a regional government document Depository Library.)

---

The following free pamphlets are available from the U.S. government's Consumer Information Center; write to S. James, Consumer Information Center, P.O. Box 100, Pueblo, CO 81002.

Enclose a $1 service fee. You can also receive a free catalog of government pamphlets by writing to the same address.

*Attract Birds*, U.S. Department of the Interior, Fish and Wildlife Service (No. 579-Z).

*Backyard Bird Feeding*, U.S. Department of the Interior, Fish and Wildlife Service (No. 580-Z).

*Homes for Birds*, U.S. Department of the Interior, Fish and Wildlife Service (No. 582-Z).

---

The list of U.S. government documents that follows is taken from the first edition (1953) of *Songbirds in Your Garden*, when it was used by the author in making his recommendations and observations about plantings for birds. Though these documents are no longer available from the Government Printing Office, you can probably find them at a university library in your state that is a U.S. government Depository Library.

*Blueberry Growing*, Farmer's Bulletin 1951.

*Growing Erect and Trailing Blackberries*, Farmer's Bulletin 1995.

*Ornamental Hedges for the Central Great Plains*, Farmer's Bulletin 2019.

*Ornamental Hedges for the Southern Great Plains*, Farmer's Bulletin 2025.

*Palm Trees in the United States*, Agricultural Information Bulletin 22.

*Raspberry Culture*, Farmer's Bulletin 887.

*Russian Olive for Wildlife and Good Land Use*, by A. E. Borell, U.S. Department of Agriculture Leaflet 292.

*Southwestern Trees: A Guide to the Native Species of New Mexico and Arizona*, U.S. Department of Agriculture Handbook No. 9.

*Useful and Ornamental Gourds*, Farmer's Bulletin 1849.

## REGIONAL GUIDES TO ORNAMENTAL PLANTING

The following booklets, pamphlets, and other sources published by state colleges of agriculture and state extension services were consulted by the author for the first edition (1953) of *Songbirds in Your Garden*. Many of them are no longer available for sale, but can be found in university libraries in the state or region they concern.

### References for the Northeast and North Central Regions

**Indiana**

*Ornamental Evergreens: Their Planting and Care*, by R. B. Hull, Extension Bulletin 320, Purdue University, Lafayette, Indiana.

**Michigan**

*Landscaping the Home Grounds*, by C. P. Halligan, Extension Bulletin 199, Michigan State College, East Lansing, Michigan.

*Hardy Shrubs for Landscape Planting in Michigan*, by C. P. Halligan, Extension Bulletin 152, Michigan State College, East Lansing, Michigan.

**New York**

*Woody Plants for Shady Places*, by R. W. Curtis and J. F. Cornman, Cornell Extension Bulletin 465, New York State College of Agriculture, Ithaca, New York.

*Shade Trees for the Home Lawn*, by Donald J. Bushey, Cornell Extension Bulletin 724, New York State College of Agriculture, Ithaca, New York.

**Ohio**

*Beautifying the Home Grounds*, by Victor H. Ries, Extension Service Bulletin 73, Ohio State University, Columbus, Ohio.

### References for the Southeast Region

**Florida**

*Propagation of Ornamental Plants*, by John V. Watkins, Extension Service Bulletin 150, University of Florida, Gainesville, Florida.

*Native and Exotic Palms of Florida*, by Harold Mowry, Extension Service Bulletin 152, University of Florida, Gainesville, Florida.

*Ornamental Hedges for Florida*, by Harold Mowry and R. D. Dickey, Extension Service Bulletin 443, University of Florida, Gainesville, Florida.

*Ground Covers for Florida Gardens*, by J. M. Crevasse, Jr., Extension Service Bulletin 473, University of Florida, Gainesville, Florida.

**Mississippi**

*An Illustrated Guide to Identification and Landscape Uses of Mississippi Native Shrubs*, by F. S. Batson, Extension Service Bulletin 369, Mississippi State University.

*An Illustrated Guide to the Care of Ornamental Trees and Shrubs*, by F. S. Batson and R. O. Monosmith, Extension Service Bulletin 35, Mississippi State University.

*An Illustrated Guide to Landscaping Mississippi Homes*, by R. O. Monosmith and F. S. Batson, Extension Service Bulletin 340, Mississippi State University.

**North Carolina**

*Planting for the Future*, by John H. Harris, Extension Circular 305, North Carolina State University, Raleigh, North Carolina.

**Tennessee**

*Better Home Grounds: Growing and Transplanting Trees and Shrubs*, by W. C. Pelton, Extension Service Publication 196, University of Tennessee, Knoxville, Tennessee.

### References for the Central Region

**Arkansas**

*Planting Materials for Arkansas Landscape Designs*, by L. H. Burton, Miscellaneous Publication No. 37, University of Arkansas, Extension Service, Fayetteville, Arkansas.

*Arbor Day*, by L. H. Burton, University of Arkansas, Extension Service, Fayetteville, Arkansas.

**Illinois**

*Selected Trees and Shrubs for Landscaping About the Illinois Home Grounds*, Extension Service Bulletin, mimeograph. Write to Agricultural Experiment Station, University of Illinois, Urbana.

*Sunflowers as a Seed and Oil Crop for Illinois*, Circular 681. Write to Agricultural Experiment Station, University of Illinois, Urbana.

**Iowa**
*Iowa Landscape Plants*, Extension Service Pamphlet, Iowa State University, Ames, Iowa.

**Louisiana**
*Louisiana Trees and Shrubs*, by Clair A. Brown, Bulletin No. 1, Louisiana Forestry Commission, Baton Rouge, Louisiana.

**Minnesota**
*Evergreens*, by L. C. Snyder *et al*, Extension Bulletin 258, University of Minnesota, St. Paul, Minnesota.

*Woody Plants for Minnesota*, by L. C. Snyder and Marvin Smith, Extension Bulletin 267, University of Minnesota, St. Paul, Minnesota.

**Missouri**
*Selection and Care of Ornamental Shrubs*, by Louise Woodruff and Julia M. Rocheford, Extension Service Circular 567, University of Missouri, Columbia, Missouri.

## References for the Plains and Desert Regions

**Kansas**
*Native Woody Plants of Kansas for Landscaping the Home Grounds*, Leaflet, Kansas State University, Manhattan, Kansas.

**Nebraska**
*Twelve Broadleaf Trees for Nebraska*, Extension Service Circular 1727, University of Nebraska, Lincoln, Nebraska.

*Developing Attractive Farmsteads in Nebraska*, Extension Service Circular 1271, University of Nebraska, Lincoln, Nebraska.

*Tree Identification Manual*, Extension Service Circular 1703, University of Nebraska, Lincoln, Nebraska.

**New Mexico**
*Ornamentals for New Mexico*, by L. C. Gibbs, Circular 224, New Mexico State University, Las Cruces, New Mexico.

*Shrubs for Northeastern New Mexico*, Bulletin 358, New Mexico State University, Las Cruces, New Mexico.

**North Dakota**
*Trees and Shrubs for Eastern North Dakota Windbreaks*, Extension Service Circular A-122, North Dakota State University, Fargo, North Dakota.

*Trees and Shrubs for Western North Dakota Windbreaks*, Extension Service Circular A-123, North Dakota State University, Fargo, North Dakota.

*Shrubs and Trees to Attract Birds*, Circular A-156, North Dakota State University, Fargo, North Dakota.

*Pruning Trees and Shrubs*, Special Circular A-63, North Dakota State University, Fargo, North Dakota.

**Oklahoma**
*Woody Plant Materials for Oklahoma*, Leaflet, Oklahoma State University, Stillwater, Oklahoma.

*Woody Plant Material*, Circular 546, Oklahoma State University, Stillwater, Oklahoma.

*Landscaping Home Grounds*, Circular 456, Oklahoma State University, Stillwater, Oklahoma.

*Home Grounds Beautification*, Circular 544, Oklahoma State University, Stillwater, Oklahoma.

**South Dakota**
*The Shade, Windbreak, and Timber Trees of South Dakota*, Bulletin 246, South Dakota State University, Brookings, South Dakota.

*Evergreens of South Dakota*, Bulletin 254, South Dakota State University, Brookings, South Dakota.

*The Ornamental Trees of South Dakota*, Bulletin 260, South Dakota State University, Brookings, South Dakota.

**Texas**
*Catalogue of the Flora of Texas*, by V. L. Cory and H. B. Parks, Extension Service Bulletin 550, Texas A & M University, College Station, Texas.

## References for the Mountain Region

**Colorado**
*Trees for Colorado Farms*, Bulletin 395-A, University of Northern Colorado, Fort Collins, Colorado.

**Idaho**
*Shrubs and Trees Noted for Their Attractive Flowers or Fruits*, Leaflet, University of Idaho, Moscow, Idaho.

**Montana**
*Woody Plant Materials for Montana*, Extension Service Leaflet, Montana

State University, Bozeman, Montana.

*Evergreens and Deciduous Trees, Shrubs*, Extension Service Leaflet, Montana State University, Bozeman, Montana.

*Vines for Landscape Use*, Extension Service Leaflet, Montana State University, Bozeman, Montana.

**Wyoming**

*Landscape Your Farm or Ranch*, Circular 104, Agricultural Extension Service, University of Wyoming, Laramie, Wyoming.

## References for the Pacific Region

**California**

*Shrubs for Coast Counties in California*, Extension Service Pamphlet, University of California, Berkeley, California.

*Trees for Southern California*, Extension Service Pamphlet, University of California, Berkeley, California.

*Trees for the Sacramento and San Joaquin Valleys of California*, Extension Service Pamphlet, University of California, Berkeley, California.

*Trees for California Coastal Districts*, Extension Service Pamphlet, University of California, Berkeley, California.

*Ornamental Gourds and Gourdlike Fruits*, Extension Service Pamphlet, University of California, Berkeley, California.

**Oregon**

*Trees to Know in Oregon*, by Charles R. Ross, Extension Bulletin 697, Oregon State University, Corvallis, Oregon.

*Culture of Trailing Berries in Oregon*, Station Bulletin 441, Oregon State University, Corvallis, Oregon.

*Raspberry Culture in Oregon*, Station Bulletin 443, Oregon State University, Corvallis, Oregon.

*Growing Small Fruits in Eastern Home Gardens*, Extension Bulletin 617, Oregon State University, Corvallis, Oregon.

## RECOMMENDED REFERENCES

### Books

American Ornithologists' Union. *Check-list of North American Birds*. 6th ed. Lawrence, Kans.: American Ornithologists' Union, 1983.

Armstrong, E. A. *Bird Display and Behaviour*. Mineola, N.Y.: Dover, 1975.

Austin, O. L., Jr. *Birds of the World*. Racine, Wis.: Western Publishing, 1961.

Baynes, E. H. *Wild Bird Guests*. New York: E. P. Dutton & Company, 1915.

Bent, A. C. *Life Histories of North American Birds*. Mineola, N.Y.: Dover, 1962-1964.

Berger, A. J. *Bird Study*. New York: John Wiley & Sons, 1971.

— *Hawaiian Bird Life*. 2nd and revised ed. Honolulu: University Press of Hawaii, 1981.

Borror, D. J. *Common Bird Songs*. Mineola, N.Y.: Dover, 1967. Audio recording.

— *Songs of Western Birds*. Mineola, N.Y.: Dover, 1971. Audio recording.

— *Birdsong and Bird Behavior*. Mineola, N.Y.: Dover, 1972. Audio recording.

Brooke, M., and T. Birkhead, eds. *The Cambridge Encyclopedia of Ornithology*. New York: Cambridge University Press, 1991.

Bull, J., ed. *A Guide to Birds of the World*. New York: Simon & Schuster, 1981.

Bull, J., E. Bull, G. Gold, and P. D. Pratt. *Birds of North America, Eastern Region*. New York: Macmillan, 1985.

Bull, J., and J. Farrand. *The Audubon Society Field Guide to North American Birds*. New York: Knopf, 1971.

Brush, A. H., and G. A. Clark, Jr., eds. *Perspectives in Ornithology*. New York: Cambridge University Press, 1983.

Burton, R. *Bird Behavior*. New York: Knopf, 1985.

Campbell, B., and E. Lack, eds. *A Dictionary of Birds*. Shipman, Va.: Buteo Books, 1985.

Collias, N. E., and E. C. Collias. *Nest Building and Bird Behavior*. Princeton, N.J.: Princeton University Press, 1984.

Cooper, J. E., and J. T. Eley. *First Aid and Care of Wild Birds*. Shipman, Va.: Buteo Books, 1979.

Cornell Laboratory of Ornithology and National Audubon Society. *Beautiful Bird Songs of the World*. Mineola, N.Y.: Dover. Audio recording.

Cornell Laboratory of Ornithology. *A Field Guide to Bird Songs of Eastern and Central North America*. Ithaca, N.Y.: Cornell Laboratory of Ornithology, 1986. Audio recording.

— *A Field Guide to Western Birds' Songs*. Ithaca, N.Y.: Cornell Laboratory of Ornithology, 1986. Audio recording.

— *Seminars in Ornithology*. Ithaca, N.Y.: Cornell Laboratory of Ornithology, 1986.

— *Watching Birds with Roger Tory Peterson*. New York: Metromedia Producers in association with Houghton Mifflin, 1986. Videocassette.

Craighead, J. J., and F. C. Craighead, Jr. *Hawks, Owls, and Wildlife*. Mineola, N.Y.: Dover, 1970.

Davison, V. E. *Attracting Birds from the Prairies to the Atlantic*. New York: Crowell, 1967.

Dennis, J. V. *A Complete Guide to Bird Feeding*. New York: Knopf, 1975.

— *The Wildlife Gardener*. New York: Knopf, 1985.

DeGraaf, Richard M. *Trees, Shrubs, and Vines for Attracting Birds: A Manual for the Northeast*. Amherst: University of Massachusetts Press, 1979.

Dorst, J. *The Migration of Birds*. Boston: Houghton Mifflin, 1963.

Farrand, J., Jr., ed. *The Audubon Society Master Guide to Birding*. 3 vols. New York: Knopf, 1983.

Fisher, J., and J. Flegg. *Watching Birds*. Shipman, Va.: Buteo Books, 1974.

Gilliard, E. T. *Living Birds of the World*. New York: Doubleday, 1958.

Godfrey, W. E. *The Birds of Canada*. Montreal: National Museum of Canada, 1966.

Gooders, J. *The Practical Ornithologist*. New York: Simon & Schuster, 1990.

Griffin, D. R. *Bird Migration*. Mineola, N.Y.: Dover, 1974.

Harrison, C. *A Field Guide to Birds' Nests, Eggs, and Nestlings of North America*. Shipman, Va.: Buteo Books, 1976.

Harrison, H. H. *A Field Guide to Birds' Nests East of the Mississippi River*. Boston: Houghton Mifflin, 1975.

— *A Field Guide to Western Birds' Nests*. Boston: Houghton Mifflin. 1979.

Hartshorne, C. *Born to Sing: An Interpretation and World Survey of Bird Songs*. Bloomington: Indiana University Press, 1973.

Hickey, J. J. *A Guide to Bird Watching*. Mineola, N.Y.: Dover, 1975.

Hochbaum, H. A. *Travels and Traditions of Waterfowl*. Minneapolis: University of Minnesota Press, 1967.

Howard, L. *Birds as Individuals*. London: Collins, 1952.

Jackson, J. A., ed. *Bird Conservation Annual*. Madison: University of Wisconsin Press, 1986.

Jellis, R. *Bird Sounds and Their Meaning*. Ithaca, N.Y.: Cornell University Press, 1984.

Johnsgard, P. A. *Hummingbirds of North America*. Washington, D.C.: Smithsonian Institution Press, 1983.

Jobling, J. A. *A Dictionary of Scientific Bird Names*. Oxford: Oxford University Press, 1991.

Kress, S. *The Audubon Society Handbook for Birders*. New York: Charles Scribner's Sons, 1981.

— *The Audubon Society Guide to Attracting Birds*. New York: Charles Scribner's Sons, 1985.

Laycock, G. *The Birdwatcher's Bible*. New York: Doubleday, 1985.

Lentz, J. E., and J. Young. *Birdwatching: A Guide for Beginners*. Santa Barbara, Calif.: Capra Press, 1985.

Lorenz, K. *King Solomon's Ring*. New York: New American Library, 1952, 1991.

Lotz, A. R. *Birding Around the World: A Guide to Observing Birds Everywhere You Travel*. New York: Dodd Mead, 1987.

Martin, A. C. *Hand-Taming Wild Birds*. Freeport, Maine: Bond Wheelwright, 1963.

Martin, A. C., H. S. Zim, and A. L. Nelson. *American Wildlife and Plants*. Mineola, N.Y.: Dover, 1951.

McClure, E. *Bird Banding*. Pacific Grove, Calif.: The Boxwood Press, 1985.

McElroy, T. P., Jr. *The Habitat Guide to Birding: A Guide to Birding East of the Rockies*. New York: Lyons and Burford, 1987.

Mead, C. *Bird Migration*. Shipman, Va.: Buteo Books, 1983.

National Geographic Society. *A Field Guide to Birds of North America*. Washington, D.C.: National Geographic Society, 1983.

— *A Guide to Bird Sounds*. Washington, D.C.: National Geographic Society, 1983.

Nice, M. M. *Studies in the Life History of the Song Sparrow*. 2 vols. Mineola, N.Y.: Dover, 1964.

Palmer, R. S., ed. *Handbook of North American Birds*. 5 vols. in multivolume series. New Haven, Conn.: Yale University Press. Vol. 1, 1962; vols. 2 and 3, 1976; vols. 4 and 5, 1988.

Pasquier, R. *Watching Birds: An Introduction to Ornithology*. Boston: Houghton Mifflin, 1985.

Peterson, R. T. *Birds over America*. New York: Dodd, Mead, 1948.

— *A Field Guide to Western Birds*. Boston: Houghton Mifflin, 1961, 1990.

— *A Field Guide to Birds of Texas and Adjacent States*. Boston: Houghton Mifflin, 1973.

— *A Field Guide to Eastern Birds: A Field Guide to Birds East of the Rockies*. Boston: Houghton Mifflin, 1984.

— *First Guide to Birds of North America*. Boston: Houghton Mifflin, 1986.

Pettingill, O. S., Jr. *Ornithology in Laboratory and Field*. New York: Academic Press, 1985.

— *My Way to Ornithology*. Norman: University of Oklahoma Press, 1992.

Piatt, J. *Adventures in Birding: Confessions of a Lister*. New York: Knopf, 1973.

Pough, R. H. *Audubon Land Bird Guide*. New York: Doubleday, 1949.

— *Audubon Water Bird Guide*. New York: Doubleday, 1957.

Robbins, C. S., B. Bruun, and H. S. Zim. *Birds of North America: A Guide to Field Identification*. Revised and expanded ed. Racine, Wis.: Western Publishing, 1983.

Scheithauer, W. *Hummingbirds*. New York: Crowell, 1967.

Schenck, M. *Your Backyard Wildlife Garden*. Emmaus, Penn.: Rodale Press, 1992.

Simpson, G. G., C. S. Pittendreigh, and L. H. Tiffany. *Life: An Introduction to Biology*. San Diego, Calif.: Harcourt Brace Jovanovich, 1957.

Skutch, A. F. *The Life of the Hummingbird*. New York: Crown, 1973, 1980.

— *Parent Birds and Their Young*. Austin: University of Texas Press, 1976.

Smith, R. L. *Ecology and Field Biology*. New York: Harper & Row, 1974.

Sparks, J. *Bird Behavior*. New York: Grosset & Dunlap, 1970.

Stefferud, A., and A. L. Nelson. *Birds in Our Lives*. Washington, D.C.: Government Printing Office, 1966.

Stokes, D. *The Bird Feeder Book: An Easy Guide to Attracting, Indentifying, and Understanding Your Feeder Birds*. Boston: Little, Brown, 1987.

— *The Bluebird Book: The Complete Guide to Attracting Bluebirds*. Boston: Little, Brown, 1991.

Stokes, D., and L. Stokes. *A Guide to Bird Behavior*. Vols. 1 and 2, Carlisle, Minn.: Stokes Nature Guides, 1979, 1983; vol. 3, Boston: Little, Brown, 1989.

Stresemann, E., and E. Mayr. *Ornithology: From Aristotle to the Present*. Cambridge, Mass.: Harvard University Press, 1975.

Temple, S. A., ed. *Bird Conservation Annual*. Madison: University of Wisconsin Press, 1983.

Terres, J. K. "My Father and the Bluebird." *The Wonders I See*. Philadelphia: Lippincott, 1960.

— *The Audubon Society Encyclopedia of North American Birds*. New York: Knopf, 1980.

— *From Laurel Hill to Siler's Bog: The Walking Adventures of a Naturalist*. Chapel Hill: University of North Carolina Press, 1993.

Thorpe, W. H. *Learning and Instinct in Animals*. London: Methuen, 1956.

Tinbergen, N. *The Herring Gull's World: A Study of the Social Behavior of Birds*. New York: Lyons and Burford, 1989.

— *The Study of Instinct*. New York: Oxford University Press, 1951, 1990.

True, D. *Hummingbirds of North America: Attracting, Feeding and Photographing*. Albuquerque: University of New Mexico Press, 1993.

Van Tyne, J., and A. J. Berger. *Fundamentals of Ornithology*. New York: John Wiley & Sons, 1976.

Wallace, G., and H. D. Mahan. *An Introduction to Ornithology*. New York: Macmillan, 1975.

Welty, J. C., and L. Baptista. *The Life of Birds*. 4th ed. Philadelphia: W. B. Saunders, 1988.

Wetmore, A. et al. *Song and Garden Birds of North America*. Washington, D.C.: National Geographic Society, 1964.

Wyman, D. *Wyman's Gardening Encyclopedia*. New York: Macmillan, 1987.

## Book Dealers

Building a basic reference library of bird books may require a long time, depending on one's resources and needs. Most books in the previous list may be ordered through one's local book store; however, in time, some have gone out of print. The following are some sellers of both new and out-of-print books. Send for their catalogues.

American Birding Association Sales
P.O. Box 6599, Colorado Springs, CO 80934-6599

Buteo Books
Route 1, Box 242, Shipman, VA 22971

Crow's Nest Book Shop
Laboratory of Ornithology, Cornell University
159 Sapsucker Woods Road, Ithaca, NY 14850

Dover Publications, Inc.
31 East Second Street, Mineola, NY 11501

Flora and Fauna Books
121 First Avenue South, Seattle, WA 98104

Donald E. Hahn
Natural History Books, Box 1004, Cottonwood, AZ 86326

Peterson Book Company
P.O. Box 966, Davenport, IA 52805

Gary Wayner Bookseller
Route 3, Box 18, Fort Payne, AL 35967-9501

## Magazines

The following are some magazines that publish articles entirely about birds, many of them about bird-attracting. They also carry advertising by dealers in bird-attracting supplies and bird-watching equipment, and by places to see birds (bird travel tours), reviews of bird books, and so on. Write to their subscription departments for information about each.

*American Birds*
950 Third Avenue, New York, NY 10022
A popular and technical publication about bird distribution

*Bird Watcher's Digest*
P.O. Box 110
Marietta, OH 45750

*Birder's World*
P.O. Box 1347, Elmhurst, IL 60126

*Living Bird Quarterly*
Laboratory of Ornithology, Cornell University
159 Sapsucker Woods Road, Ithaca, NY 14850

*Nature Society News*
Purple Martin Junction, Griggsville, IL 62340

## Technical Ornithological Journals

*The Auk: A Quarterly Journal of Ornithology*
Ornithological Societies of North America
P.O. Box 1897
Lawrence, KS 66044-8879
Official publication of the American Ornithologists' Union

*The Condor: Journal of the Cooper Ornithological Society*
Ornithological Societies of North America
(See address under *The Auk*)

*Journal of Field Ornithology: A Journal of Ornithological Investigation*
(Formerly *Bird-Banding*)
Northeastern Bird-Banding Association
c/o Allen Press
P.O. Box 368
Lawrence, KS 66044

*The Wilson Bulletin*
Ornithological Societies of North America
(See address under *The Auk*)
Official publication of the Wilson Ornithological Society

# *Index*

*Note: Page numbers followed by the letter f indicates a figure; page numbers followed by the letter t indicate a table.*